MATT RIDLEY's books on genetics, evolution and society have been translated into thirty languages. He writes regularly for *The Times* and the *Wall Street Journal* and is a member of the House of Lords. He lives in Northumberland.

Praise for *The Evolution of Everything*:

'A highly intelligent and bracingly iconoclastic view of the world. It forces us to see life through new eyes'
New York Times Book Review

'He argues we live in a bottom-up world . . . a compelling argument and in this fascinating work, an evolution from Ridley's other books, such as *The Rational Optimist* or *The Origins of Virtue*, he takes it to all realms of knowledge and how new ideas emerge . . . Ridley has amassed such a weight of fascinating evidence and anecdote that the pages fly by'
The Times

'This is a book of remarkable scope (when Ridley says everything, he isn't exaggerating), clearly written by a polymath who reads whatever is interesting, old and new. What's more, it does not have the feel of a book written on commission so much as one that has been slowly assembling its own emergent thesis over time, tentatively testing and sometimes rejecting ideas along the way. As so often in nature, something wonderful has thereby come about'
Literary Review

'The book displays his wide and deep knowledge of many different fields. It is fast paced and elegantly written. Few readers will come away without fresh information and a challenge to their preconceptions'
Prospect

'Readable, provocative and infuriating'
New Statesman

'This penetrating book is Mr. Ridley's best and most important work to date . . . there is something profoundly democratic and egalitarian – even anti-elitist – in this bottom-up approach: Everyone can have a role in bringing about change'
Wall Street Journal

By the same author

The Red Queen:
Sex and the Evolution of Human Nature

The Origins of Virtue

Genome:
The Autobiography of a Species in 23 Chapters

Nature via Nurture:
Genes, Experience and What Makes Us Human

Francis Crick: Discoverer of the Genetic Code

The Rational Optimist: How Prosperity Evolves

Matt Ridley

THE EVOLUTION OF EVERYTHING

How Small Changes Transform Our World

4th ESTATE · London

4th Estate
An imprint of HarperCollinsPublishers
1 London Bridge Street
London SE1 9GF
www.4thestate.co.uk

First published in Great Britain in 2015 by 4th Estate
This 4th Estate paperback edition first published in 2016

ISBN 978 0 00 754247 5

Typeset in Sabon by Birdy Book Design

Printed and bound in Great Britain by Clays Ltd, St Ives plc

MIX
Paper from
responsible sources
FSC™ C007454

CONTENTS

Prologue: The General Theory of Evolution   1

1   The Evolution of the Universe   7

2   The Evolution of Morality   21

3   The Evolution of Life   37

4   The Evolution of Genes   59

5   The Evolution of Culture   76

6   The Evolution of the Economy   96

7   The Evolution of Technology   118

8   The Evolution of the Mind   140

9   The Evolution of Personality   155

10   The Evolution of Education   174

11   The Evolution of Population   193

12   The Evolution of Leadership   215

13   The Evolution of Government   235

14   The Evolution of Religion   256

15   The Evolution of Money   277

16   The Evolution of the Internet   299

Epilogue: The Evolution of the Future   317

Acknowledgements   321

Sources and Further Reading   323

Index   343

PROLOGUE

The General Theory of Evolution

The word 'evolution' originally means 'unfolding'. Evolution is a story, a narrative of how things change. It is a word freighted with many other meanings, of particular kinds of change. It implies the emergence of something from something else. It has come to carry a connotation of incremental and gradual change, the opposite of sudden revolution. It is both spontaneous and inexorable. It suggests cumulative change from simple beginnings. It brings the implication of change that comes from within, rather than being directed from without. It also usually implies change that has no goal, but is open-minded about where it ends up. And it has of course acquired the very specific meaning of genetic descent with modification over the generations in biological creatures through the mechanism of natural selection.

This book argues that evolution is happening all around us. It is the best way of understanding how the human world changes, as well as the natural world. Change in human institutions, artefacts and habits is incremental, inexorable and inevitable. It follows a narrative, going from one stage to the next; it creeps rather than jumps; it has its own spontaneous momentum, rather than being driven from outside; it has no goal or end in mind; and it largely happens by trial and error – a version of natural selection. Take, for example, electric light. When an obscure engineer named Thomas Newcomen in 1712 hit upon the first practical method of turning heat into work, he could have had no notion that the basic principle behind his invention – the expansion of water when boiled to make steam – would eventually result, via innumerable small steps, in machines that generate electricity to provide artificial light: heat to work to light. The change from incandescent to fluorescent and next to LED light is still unfolding. The sequence of events was and is evolutionary.

My argument will be that in all these senses, evolution is far more common, and far more influential, than most people recognise. It is not confined to genetic systems, but explains the way that virtually all of human culture changes: from morality to technology, from money to religion. The way in which these streams of human culture flow is gradual, incremental, undirected, emergent and driven by natural selection among competing ideas. People are the victims, more often than the perpetrators, of unintended change. And though it has no goal in mind, cultural evolution none the less produces functional and ingenious solutions to problems – what biologists call adaptation. In the case of the forms and behaviours of animals and plants, we find this apparent purposefulness hard to explain without imputing deliberate design. How can it not be that the eye was designed for seeing? In the same way, we assume that when we find human culture being well adapted to solve human problems, we tend to assume that this is because some clever person designed it with that end in mind. So we tend to give too much credit to whichever clever person is standing nearby at the right moment.

The way that human history is taught can therefore mislead, because it places far too much emphasis on design, direction and planning, and far too little on evolution. Thus, it seems that generals win battles; politicians run countries; scientists discover truths; artists create genres; inventors make breakthroughs; teachers shape minds; philosophers change minds; priests teach morality; businessmen lead businesses; conspirators cause crises;

gods make morality. Not just individuals, but institutions too: Goldman Sachs, the Communist Party, the Catholic Church, Al Qaeda – these are said to shape the world.

That's the way I was taught. I now think it is more often wrong than right. Individuals can make a difference, of course, and so can political parties or big companies. Leadership still matters. But if there is one dominant myth about the world, one huge mistake we all make, one blind spot, it is that we all go around assuming the world is much more of a planned place than it is. As a result, again and again we mistake cause for effect; we blame the sailing boat for the wind, or credit the by-stander with causing the event. A battle is won, so a general must have won it (not the malaria epidemic that debilitated the enemy army); a child learns, so a teacher must have taught her (not the books, peers and curiosity that the teacher helped her find); a species is saved, so a conservationist must have saved it (not the invention of fertiliser which cut the amount of land needed to feed the population); an invention is made, so an inventor must have invented it (not the inexorable, inevitable ripeness of the next technological step); a crisis occurs, so we see a con-spiracy (and not a cock-up). We describe the world as if people and institutions were always in charge, when often they are not. As Nassim Taleb remarks in his book *Antifragile*, in a complex world the very notion of 'cause' is suspect: 'another reason to ignore newspapers with their constant supply of causes for things'.

Taleb is brutally dismissive of what he mockingly calls the Soviet-Harvard illusion, which he defines as lecturing birds on flight and thinking that the lecture caused their skill at flying. Adam Smith was no less rude about what he called the man of system, who imagines 'that he can arrange the different members of a great society with as much ease as the hand arranges the different pieces upon a chess-board', without considering that on the great chessboard of human society, the pieces have a motion of their own.

To use a word coined by Abraham Lincoln, I hope gradually to 'disenthrall' you over the course of this book, from the obsession with human intentionality, design and planning. I want to do for every aspect of the human world a little bit of what Charles Darwin did for biology, and get you to see past the illusion of design, to see the emergent, unplanned, inexorable and beautiful process of change that lies underneath.

I have often noticed that human beings are surprisingly bad at explaining their own world. If an anthropologist from Alpha Centauri were to arrive here and ask some penetrating questions, he would get no good answers. Why is the homicide rate falling all around the world? Criminologists cannot agree. Why is global average income more than ten times what it was in the nineteenth century? Economic historians are divided. Why did some Africans start to invent cumulative technology and civilisation around 200,000 years ago? Anthropologists do not know. How does the world economy work? Economists pretend to explain, but they cannot really do so in any detail.

These phenomena belong in a strange category, first defined in 1767 by a Scottish army chaplain by the name of Adam Ferguson: they are the result of human action, but not of human design. They are *evolutionary* phenomena, in the original meaning of the word – they unfold. And evolutionary phenomena such as these are everywhere and in everything. Yet we fail to recognise this category. Our language and our thought divide the world into two kinds of things – those designed and made by people, and natural phenomena with no order or function. The economist Russ Roberts once pointed out that we have no word to encompass such phenomena. The umbrella that keeps you dry in a shower of rain is the result of both human action and human design, whereas the rainstorm that soaks you when you forget it is neither. But what about the system that enables a local shop to sell you an umbrella, or the word umbrella itself, or the etiquette that demands that you tilt your umbrella to one side to let another pedestrian pass? These – markets, language, customs – are man-made things. But none of them is designed by a human being. They all emerged unplanned.

We transfer this thinking back into our understanding of the natural world too. We see purposeful design in nature, rather than emergent evolution. We look for hierarchy in the genome, for a 'self' in the brain, and for free will in the mind. We latch on to any excuse to blame an extreme weather event on human agency – whether witchdoctoring or man-made global warming.

Far more than we like to admit, the world is to a remarkable extent a self-organising, self-changing place. Patterns emerge, trends evolve. Skeins of geese form Vs in the sky without meaning to, termites build cathedrals without architects, bees make hexagonal honeycombs without instruction, brains take shape without brain-makers, learning can happen without teaching, political events are shaped by history rather than vice versa. The genome has no master gene, the brain has no command centre, the English language has no director, the economy has no chief executive, society has no president, the common law has no chief justice, the climate has no control knob, history has no five-star general.

In society, people are the victims and even the immediate agents of change, but more often than not the causes are elsewhere – they are emergent, collective, inexorable forces. The most powerful of these inexorable forces is biological evolution by natural selection itself, but there are other, simpler forms of evolutionary, unplanned change. Indeed, to borrow a phrase from a theorist of innovation, Richard Webb, Darwinism is the 'special theory of evolution'; there's a general theory of evolution too, and it applies to much more than biology. It applies to society, money, technology, language, law, culture, music, violence, history, education, politics, God, morality. The general theory says that things do not stay the same; they change gradually but inexorably; they show 'path dependence'; they show descent with modification; they show trial and error; they show selective persistence. And human beings none the less take credit for this process of endogenous change as if it was directed from above.

This truth continues to elude most intellectuals on the left as well as the right, who remain in effect 'creationists'. The obsession with which those on the right resist Charles Darwin's insight – that the complexity of nature does not imply a designer – matches the obsession with which those on the left resist Adam Smith's insight – that the complexity of society does not imply a planner. In the pages that follow, I shall take on this creationism in all its forms.

1

The Evolution of the Universe

If you possess a firm grasp of these tenets, you will see
That Nature, rid of harsh taskmasters, all at once is free
And everything she does, does on her own, so that gods play
No part . . .

Lucretius, *De Rerum Natura*, Book 2, lines 1090–3

A 'skyhook' is an imaginary device for hanging an object from
the sky. The word originated in a sarcastic remark by a frustrated
pilot of a reconnaissance plane in the First World War, when told
to stay in the same place for an hour: 'This machine is not fitted
with skyhooks,' he replied. The philosopher Daniel Dennett used
the skyhook as a metaphor for the argument that life shows evidence of an intelligent designer. He contrasted skyhooks with
cranes – the first impose a solution, explanation or plan on the
world from on high; the second allow solutions, explanations
or patterns to emerge from the ground up, as natural selection
does.

The history of Western thought is dominated by skyhooks, by
devices for explaining the world as the outcome of design and
planning. Plato said that society worked by imitating a designed
cosmic order, a belief in which should be coercively enforced.

Aristotle said that you should look for inherent principles of intentionality and development – souls – within matter. Homer said gods decided the outcome of battles. St Paul said that you should behave morally because Jesus told you so. Mohamed said you should obey God's word as transmitted through the Koran. Luther said that your fate was in God's hands. Hobbes said that social order came from a monarch, or what he called 'Leviathan' – the state. Kant said morality transcended human experience. Nietzsche said that strong leaders made for good societies. Marx said that the state was the means of delivering economic and social progress. Again and again, we have told ourselves that there is a top–down description of the world, and a top–down prescription by which we should live.

But there is another stream of thought that has tried and usually failed to break through. Perhaps its earliest exponent was Epicurus, a Greek philosopher about whom we know very little. From what later writers said about his writings, we know that he was born in 341 BC and thought (as far as we can tell) that the physical world, the living world, human society and the morality by which we live all emerged as spontaneous phenomena, requiring no divine intervention nor a benign monarch or nanny state to explain them. As interpreted by his followers, Epicurus believed, following another Greek philosopher, Democritus, that the world consisted not of lots of special substances including spirits and humours, but simply of two kinds of thing: voids and atoms. Everything, said Epicurus, is made of invisibly small and indestructible atoms, separated by voids; the atoms obey the laws of nature and every phenomenon is the result of natural causes. This was a startlingly prescient conclusion for the fourth century BC.

Unfortunately Epicurus's writings did not survive. But three hundred years later, his ideas were revived and explored in a lengthy, eloquent and unfinished poem, *De Rerum Natura* (*Of the Nature of Things*), by the Roman poet Titus Lucretius

Carus, who probably died in mid-stanza around 49 BC, just as dictatorship was looming in Rome. Around this time, in Gustave Flaubert's words, 'when the gods had ceased to be, and Christ had not yet come, there was a unique moment in history, between Cicero and Marcus Aurelius when man stood alone'. Exaggerated maybe, but free thinking was at least more possible then than before or after. Lucretius was more subversive, open-minded and far-seeing than either of those politicians (Cicero admired, but disagreed with, him). His poem rejects all magic, mysticism, superstition, religion and myth. It sticks to an un-alloyed empiricism.

As the Harvard historian Stephen Greenblatt has docu-mented, a bald list of the propositions Lucretius advances in the unfinished 7,400 hexameters of *De Rerum Natura* could serve as an agenda for modernity. He anticipated modern physics by arguing that everything is made of different combinations of a limited set of invisible particles, moving in a void. He grasped the current idea that the universe has no creator, Providence is a fantasy and there is no end or purpose to existence, only ceaseless creation and destruction, governed entirely by chance. He foreshadowed Darwin in suggesting that nature ceaselessly experiments, and those creatures that can adapt and reproduce will thrive. He was with modern philosophers and historians in suggesting that the universe was not created for or about human beings, that we are not special, and there was no Golden Age of tranquillity and plenty in the distant past, but only a primitive battle for survival. He was like modern atheists in arguing that the soul dies, there is no afterlife, all organised religions are superstitious delusions and invariably cruel, and angels, demons or ghosts do not exist. In his ethics he thought the highest goal of human life is the enhancement of pleasure and the reduction of pain.

Thanks largely to Greenblatt's marvellous book *The Swerve*, I have only recently come to know Lucretius, and to appre-ciate the extent to which I am, and always have been without knowing it, a Lucretian/Epicurean. Reading his poem in A.E. Stallings's beautiful translation in my sixth decade is to be left fuming at my educators. How could they have made me waste all those years at school plodding through the tedious platitudes and pedestrian prose of Jesus Christ or Julius Caesar, when they could have been telling me about Lucretius instead, or as well? Even Virgil was writing partly in reaction to Lucretius, keen to re-establish respect for gods, rulers and top–down ideas in general. Lucretius's notion of the ceaseless mutation of forms composed of indestructible substances – which the Spanish-born philosopher George Santayana called the greatest thought that mankind has ever hit upon – has been one of the persistent themes of my own writing. It is the central idea behind not just physics and chemistry, but evolution, ecology and economics too. Had the Christians not suppressed Lucretius, we would surely have discovered Darwinism centuries before we did.

The Lucretian heresy

It is by the thinnest of threads that we even know the poem *De Rerum Natura*. Although it was mentioned and celebrated by contemporaries, and charred fragments of it have been found in the Villa of the Papyri at Herculaneum (a library belonging probably to Julius Caesar's father-in-law), it sank into obscurity for much of history. Passing quotations from it in the ninth century AD show that it was very occasionally being read by monks, but by 1417 no copy had been in wide circulation among scholars for more than a millennium. As a text it was effectively extinct. Why?

It is not hard to answer that question. Lucretius's special contempt for all forms of superstition, and indeed his atomism, which contradicted the doctrine of transubstantiation, condemned him to obscurity once the Christians took charge. His elevation of the pleasure principle – that the pursuit of pleasure could lead to goodness and that there was nothing nice about pain – was incompatible with the recurring Christian obsession that pleasure is sinful and suffering virtuous.*

Whereas Plato and Aristotle could be accommodated within Christianity, because of their belief in the immortality of the soul and the evidence for design, the Epicurean heresy was so threatening to the Christian Church that Lucretius had to be suppressed. His atheism is explicit, even Dawkinsian, in its directness. The historian of philosophy Anthony Gottlieb compares a passage from Lucretius with one from Richard Dawkins's *The Selfish Gene*. The first talks of 'the generation of living creatures' by 'every sort of combination and motion'; the second of how 'unordered atoms could group themselves into ever more complex patterns until they ended up manufacturing people'. Lucretius was, carped John Dryden, at times 'so much an atheist, he forgot to be a poet'. He talks about people 'crushed beneath the weight of superstition', claims that 'it is religion breeds wickedness' and aims to give us 'the power to fight against the superstitions and the threats of priests'. Little wonder they tried to stamp him out.

They almost succeeded. St Jerome – keen to illustrate the wages of sin – dismissed Lucretius as a lunatic, driven mad by a love potion, who then committed suicide. No evidence to support these calumnies exists; saints do not show their sources. The charge that all Epicureans were scandalous hedonists was trumped up and spread abroad, and it persists to this day. Copies of the poem were rooted out of libraries and destroyed, as were

---

* Greenblatt's book has been severely criticised, as successful books often are, by other academics, mainly on the grounds that he stands accused of exaggerating the illiteracy and ignorance of the medieval clerisy, that he misses the fact that the poem was at least sporadically mentioned in the ninth century, and that he is too harsh towards religious thinking. But in his main argument that *De Rerum Natura* was suppressed and attacked by Christianity – even after its rediscovery – and had an influence on the Renaissance and the Enlightenment, once it was widely circulated after 1417, there is no doubt that Greenblatt is right.

any other Epicurean and sceptical works. Almost all traces of such materialist and humanist thought had apparently long since vanished from Europe when in 1417 a Florentine scholar and recently unemployed papal secretary named Gian Francesco Poggio Bracciolini, stumbled upon a copy of the whole poem. Poggio was hunting for rare manuscripts in libraries in central Germany when he came across a copy of *De Rerum Natura* in a monastic library (probably at Fulda). He sent a hastily-made copy to his wealthy bibliophile friend Niccolò Niccoli, whose transcription was then copied more than fifty times. In 1473 the book was printed and the Lucretian heresy began to infect minds all across Europe.

Newton's nudge

In his passionate attachment to rationalism, materialism, naturalism, humanism and liberty, Lucretius deserves a special place in the history of Western thought, even above the beauty of his poetry. The Renaissance, the scientific revolution, the Enlightenment and the American Revolution were all inspired by people who had to some degree imbibed Lucretius. Botticelli's *Venus* effectively depicts the opening scene of Lucretius's poem. Giordano Bruno went to the stake, with his mouth pinned shut to silence his heresies, for quoting Lucretius on the recombination of atoms and the awe with which we should embrace the idea that human beings are not the purpose of the universe. Galileo's Lucretian atomism, as well as his Copernican heliocentrism, was used against him at his trial. Indeed, the historian of science Catherine Wilson has argued that the whole of seventeenth-century empiricism, started by Pierre Gassendi in opposition to Descartes, and taken up by the most influential thinkers of the age, including Thomas Hobbes, Robert Boyle, John Locke, Gottfried Leibniz and Bishop Berkeley, was fuelled to a remarkable extent by the sudden popularity of Lucretius.

As Lucretian ideas percolated, the physicists were the first to see where they led. Isaac Newton became acquainted with Epicurean atomism as a student at Cambridge, when he read a book by Walter Charleton expounding Gassendi's interpretation of Lucretius. Later he acquired a Latin edition of *De Rerum Natura* itself, which survives from his library and shows signs of heavy use. He echoed Lucretian ideas about voids between atoms throughout his books, especially the *Opticks*.

Newton was by no means the first modern thinker to banish a skyhook, but he was one of the best. He explained the orbits of the planets and the falling of apples by gravity, not God. In doing so, he did away with the need for perpetual divine interference and supervision by an overworked creator. Gravity kept the earth orbiting the sun without having to be told. Jehovah might have kicked the ball, but it rolled down the hill of its own accord.

Yet Newton's disenthralment was distinctly limited. He was furious with anybody who read into this that God might not be in ultimate charge, let alone not exist. He asserted firmly that: 'This most elegant system of the sun, planets, and comets could not have arisen without the design and dominion of an intelligent and powerful being.' His reasoning was that, according to his calculations, the solar system would eventually spin off into chaos. Since it apparently did not, God must be intervening periodically to nudge the planets back into their orbits. Jehovah has a job after all, just a part-time one.

The swerve

That's that then. A skyhook still exists, just out of sight. Again and again this was the pattern of the Enlightenment: gain a yard of ground from God, but then insist he still holds the field beyond and always will. It did not matter how many skyhooks were found to be illusory, the next one was always going to prove real. Indeed, so common is the habit of suddenly seeing design, after all the hard work has been done to show that emergence is more plausible, that I shall borrow a name for it – the swerve. Lucretius himself was the first to swerve. In a world composed of atoms whose motions were predictable, Lucretius (channelling Democritus and Epicurus) could not explain the apparent human capacity for free will. In order to do so, he suggested, arbitrarily, that atoms must occasionally swerve unpredictably, because the gods make them do so. This failure of nerve on the part of the poet has been known since as the Lucretian swerve, but I intend to use the same phrase more generally for each occasion on which I catch a philosopher swerving to explain something he struggles to understand, and positing an arbitrary skyhook. Watch out, in the pages that follow, for many Lucretian swerves.

Newton's rival, Gottfried Leibniz, in his 1710 treatise on theodicy, attempted a sort of mathematical proof that God existed. Evil stalked the world, he concluded, the better to bring out the best in people. God was always calculating carefully how to minimise evil, if necessary by allowing disasters to occur that killed more bad people than good. Voltaire mocked Leibniz's 'optimism', a word that then meant almost the opposite of what it means today: that the world was perfect and unimprovable ('optimal'), because God had made it. After 60,000 people died in the Lisbon earthquake of 1755, on the morning of All Saints' Day when the churches were full, theologians followed Leibniz in explaining helpfully that Lisbon had earned its punishment by sinning. This was too much for Voltaire, who asked sardonically in a poem: 'Was then more vice in fallen Lisbon found/Than Paris, where voluptuous joys abound?'

Newton's French follower Pierre-Louis Maupertuis went to Swedish Lapland to prove that the earth was flattened towards the poles, as Newtonian mechanics predicted. He then moved on from Newton by rejecting other arguments for the existence of God founded on the wonders of nature, or the regularity of the solar system. But having gone thus far, he suddenly stopped (his Lucretian swerve), concluding that his own 'least action' principle to explain motion displayed such wisdom on the part of nature that it must be the product of a wise creator. Or, to paraphrase Maupertuis, if God's as clever as me, he must exist. A blazing non sequitur.

Voltaire, perhaps irritated by the fact that his mathematically gifted mistress Emilie, Marquise du Châtelet had slept with Maupertuis and had written in defence of Leibniz, then based his character Dr Pangloss in his novel *Candide* on an amalgam of Leibniz and Maupertuis. Pangloss remains blissfully persuaded – and convinces the naïve Candide – that this is the best of all possible worlds, even as they both experience syphilis, shipwreck, earthquake, fire, slavery and being hanged. Voltaire's contempt for theodicy derived directly and explicitly from Lucretius, whose arguments he borrowed throughout life, styling himself at one point the 'latter-day Lucretius'.

Pasta or worms?

Voltaire was by no means the first poet or prose stylist to draw upon Lucretius, nor would he be the last. Thomas More tried to reconcile Lucretian pleasure with faith in *Utopia*. Montaigne quoted Lucretius frequently, and echoed him in saying 'the world is but a perennial movement . . . all things in it are in constant motion'; he recommended that we 'fall back into Epicurus' infinity of atoms'. Britain's Elizabethan and Jacobean poets, including Edmund Spenser, William Shakespeare, John Donne and Francis Bacon, all play with themes of explicit materialism and atomism that came either directly or indirectly from Lucretius. Ben Jonson heavily annotated his Dutch edition of Lucretius. Machiavelli copied out *De Rerum Natura* in his youth. Molière, Dryden and John Evelyn translated it; John Milton and Alexander Pope emulated, echoed and attempted to rebut it.

Thomas Jefferson, who collected five Latin versions of *De Rerum Natura* along with translations into three languages, declared himself an Epicurean, and perhaps deliberately echoed Lucretius in his phrase 'the pursuit of happiness'. The poet and physician Erasmus Darwin, who helped inspire not just his evolutionary grandson but many of the Romantic poets too, wrote his epic, erotic, evolutionary, philosophical poems in conscious imitation of Lucretius. His last poem, *The Temple of Nature*, was intended as his version of *De Rerum Natura*.

The influence of this great Roman materialist culminates rather neatly in the moment when Mary Shelley had the idea for *Frankenstein*. She had her epiphany after listening to her husband Percy discuss with George, Lord Byron, the coming alive of 'vermicelli' that had been left to ferment, in experiments of 'Dr Darwin'. Given that Shelley, Byron and Erasmus Darwin were all enthusiastic Lucretians, perhaps she misheard and, rather than debating the resurrection of pasta, they were actually quoting the passage in *De Rerum Natura* (and Darwin's experimental imitation of it) where Lucretius discusses spontaneous generation of little worms in rotting vegetable matter – 'vermiculos'. Here is the history of Western thought in a single incident: a Classical writer, rediscovered in the Renaissance, who inspired the Enlightenment and influenced the Romantic movement, then sparks the most famous Gothic novel, whose villain becomes a recurring star of modern cinema.

Lucretius haunted philosophers of the Enlightenment, daring free thinkers further down the path that leads away from creationist thinking. Pierre Bayle, in his *Thoughts on the Comet of 1680*, closely followed Lucretius's Book 5 in suggesting that the power of religion derived from fear. Montesquieu channelled Lucretius in the very first sentence of *The Spirit of the Laws* (1748): 'Laws in their most general signification, are the necessary relations arising from the *nature of things*' (my emphasis). Denis Diderot in his *Philosophical Thoughts* echoed Lucretius to the effect that nature was devoid of purpose, the motto for his book being a line from *De Rerum Natura*: 'Now we see out of the dark what is in the light'. Later, in *The Letter on the Blind and the Deaf*, Diderot suggested that God himself was a mere product of the senses, and went to jail for the heresy. The atheist philosopher Paul-

Henri, baron d'Holbach, took Lucretian ideas to their ultimate extreme in his *Le Système de la Nature* of 1770. D'Holbach saw nothing but cause and effect, and matter in motion: 'no necessity to have recourse to supernatural powers to account for the formation of things'.

One place where such scepticism began to take hold was in geology. James Hutton, a farmer from southern Scotland, in 1785 laid out a theory that the rocks beneath our feet were made by processes of erosion and uplift that are still at work today, and that no great Noachian flood was needed to explain seashells on mountaintops: 'Hence we are led to conclude, that the greater part of our land, if not the whole, had been produced by operations natural to this globe.' He glimpsed the vast depths of geological time, saying famously, 'We find no vestige of a beginning – no prospect of an end.' For this he was vilified as a blasphemer and an atheist. The leading Irish scientist Richard Kirwan even went as far as to hint that ideas like Hutton's contributed to dangerous events like the French Revolution, remarking on how they had 'proved too favourable to the structure of various systems of atheism or infidelity, as these have been in their turn to turbulence and immorality'.

No need of that hypothesis

The physicists, who had set the pace in tearing down sky-hooks, continued to surprise the world. It fell to Pierre-Simon Laplace (using Emilie du Châtelet's improvements to cumbersome Newtonian geometry) to take Newtonism to its logical conclusion. Laplace argued that the present state of the universe was 'the effect of its past and the cause of its future'. If an intellect were powerful enough to calculate every effect of every cause, then 'nothing would be uncertain and the future just like the past would be present before its eyes'. By mathematically showing that there was no need in the astronomical world even for Newton's Nudge God to intervene to keep the solar system stable, Laplace took away that skyhook. 'I had no need of that hypothesis,' he told Napoleon.

The certainty of Laplace's determinism eventually crumbled in the twentieth century under assault from two directions – quantum mechanics and chaos theory. At the subatomic level, the world turned out to be very far from Newtonian, with uncertainty built into the very fabric of matter. Even at the astronomical scale, Henri Poincaré discovered that some arrangements of heavenly bodies resulted in perpetual instability. And as the meteorologist Edward Lorenz realised, exquisite sensitivity to initial conditions meant that weather systems were inherently unpredictable, asking, famously, in the title of a lecture in 1972: 'Does the flap of a butterfly's wings in Brazil set off a tornado in Texas?'

But here's the thing. These assaults on determinism came from below, not above; from within, not without. If anything they made the world a still more Lucretian place. The impossibility of forecasting the position of an electron, or the weather a year ahead, made the world proof against the confidence of prognosticators and experts and planners.

The puddle that fits its pothole

Briefly in the late twentieth century, some astronomers bought into a new skyhook called the 'anthropic principle'. In various forms, this argued that the conditions of the universe, and the particular values of certain parameters, seemed ideally suited to the emergence of life. In other words, if things had been just a little bit different, then stable suns, watery worlds and polymerised carbon would not be possible, so life could never get started. This stroke of cosmic luck implied that we lived in some kind of privileged universe uncannily suitable for us, and this was somehow spooky and cool.

Certainly, there do seem to be some remarkably fortuitous features of our own universe without which life would be impossible. If the cosmological constant were any larger, the pressure of antigravity would be greater and the universe would have blown itself to smithereens long before galaxies, stars and planets could have evolved. Electrical and nuclear forces are just the right strength for carbon to be one of the most common elements, and carbon is vital to life because of its capacity to form multiple bonds. Molecular bonds are just the right strength to be stable but breakable at the sort of temperatures found at the typical distance of a planet from a star: any weaker and the universe would be too hot for chemistry, any stronger and it would be too cold.

True, but to anybody outside a small clique of cosmologists who had spent too long with their telescopes, the idea of the anthropic principle was either banal or barmy, depending on how seriously you take it. It so obviously confuses cause and effect. Life adapted to the laws of physics, not vice versa. In a world where water is liquid, carbon can polymerise and solar systems last for billions of years, then life emerged as a carbon-based system with water-soluble proteins in fluid-filled cells. In a different world, a different kind of life might emerge, if it could. As David Waltham puts it in his book *Lucky Planet*, 'It is all but inevitable that we occupy a favoured location, one of the rare neighbourhoods where by-laws allow the emergence of intelligent life.' No anthropic principle needed.

Waltham himself goes on to make the argument that the earth may be rare or even unique because of the string of ridiculous coincidences required to produce a planet with a stable temperature with liquid water on it for four billion years. The moon was a particular stroke of luck, having been formed by an interplanetary collision and having then withdrawn slowly into space as a result of the earth's tides (it is now ten times as far away as when it first formed). Had the moon been a tiny bit bigger or smaller, and the earth's day a tiny bit longer or shorter after the collision, then we would have had an unstable axis and a tendency to periodic life-destroying climate catastrophes that would have precluded the emergence of intelligent life. God might claim credit for this lunar coincidence, but Gaia – James Lovelock's theory that life itself controls the climate – cannot. So we may be extraordinarily lucky and vanishingly rare. But that does not make us special: we would not be here if it had not worked out so far.

Leave the last word on the anthropic principle to Douglas Adams: 'Imagine a puddle waking up one morning and thinking, "This is an interesting world I find myself in – an interesting hole I find myself in – fits me rather neatly, doesn't it? In fact it fits me staggeringly well, may have been made to have me in it!" '

Thinking for ourselves

It is no accident that political and economic enlightenment came in the wake of Newton and his followers. As David Bodanis argues in his biography of Voltaire and his mistress, *Passionate Minds*, people would be inspired by Newton's example to question traditions around them that had apparently been accepted since time immemorial. 'Authority no longer had to come from what you were told by a priest or a royal official, and the whole establishment of the established church or the state behind them. It could come, dangerously, from small, portable books – and even from ideas you came to yourself.'

Gradually, by reading Lucretius and by experiment and thought, the Enlightenment embraced the idea that you could explain astronomy, biology and society without recourse to intelligent design. Nikolaus Copernicus, Galileo Galilei, Baruch Spinoza and Isaac Newton made their tentative steps away from top–down thinking and into the bottom–up world. Then, with gathering excitement, Locke and Montesquieu, Voltaire and Diderot, Hume and Smith, Franklin and Jefferson, Darwin and Wallace, would commit similar heresies against design. Natural explanations displaced supernatural ones. The emergent world emerged.

2

The Evolution of Morality

O miserable minds of men! O hearts that cannot see!
Beset by such great dangers and in such obscurity
You spend your lot of life! Don't you know it's plain
That all your nature yelps for is a body free from pain,
And, to enjoy pleasure, a mind removed from fear and care?

Lucretius, *De Rerum Natura*, Book 2, lines 1–5

Soon a far more subversive thought evolved from the followers of Lucretius and Newton. What if morality itself was not handed down from the Judeo-Christian God as a prescription? And was not even the imitation of a Platonic ideal, but was a spontaneous thing produced by social interaction among people seeking to find ways to get along? In 1689, John Locke argued for religious tolerance – though not for atheists or Catholics – and brought a storm of protest down upon his head from those who saw government enforcement of religious orthodoxy as the only thing that prevented society from descending into chaos. But the idea of spontaneous morality did not die out, and some time later David Hume and then Adam Smith began to dust it off and show it to the world: morality as a spontaneous phenomenon. Hume realised that it was good for society if people were nice to each other, so he thought that rational calculation, rather than moral instruction, lay behind social cohesion. Smith went one step further, and suggested that morality emerged unbidden and unplanned from a peculiar feature of human nature: sympathy.

Quite how a shy, awkward, unmarried professor from Kirkcaldy who lived with his mother and ended his life as a customs inspector came to have such piercing insights into human nature is one of history's great mysteries. But Adam Smith was lucky in his friends. Being taught by the brilliant Irish lecturer Francis Hutcheson, talking regularly with David Hume, and reading Denis Diderot's new *Encyclopédie*, with its relentless interest in bottom–up explanations, gave him plenty with which to get started. At Balliol College, Oxford, he found the lecturers 'had altogether given up even the pretence of teaching', but the library was 'marvellous'. Teaching in Glasgow gave him experience of merchants in a thriving trading port and 'a feudal, Calvinist world dissolving into a commercial, capitalist one'. Glasgow had seen explosive growth thanks to increasing trade with the New World in the eighteenth century, and was fizzing with entrepreneurial energy. Later, floating around France as the tutor to the young Duke of Buccleuch enabled Smith to meet d'Holbach and Voltaire, who thought him 'an excellent man. We have nothing to compare with him.' But that was after his first, penetrating book on human nature and the evolution of morality. Anyway, somehow this shy Scottish man stumbled upon the insights to explore two gigantic ideas that were far ahead of their time. Both concerned emergent, evolutionary phenomena: things that are the result of human action, but not the result of human design.

Adam Smith spent his life exploring and explaining such emergent phenomena, beginning with language and morality, moving on to markets and the economy, ending with the law, though he never published his planned book on jurisprudence. Smith began lecturing on moral philosophy at Glasgow University in the 1750s, and in 1759 he put together his lectures as a book, *The Theory of Moral Sentiments*. Today it seems nothing remarkable: a dense and verbose eighteenth-century ramble through ideas about ethics. It is not a rattling read. But in its time it was surely one of the most subversive books ever written. Remember that morality was something that you had to be taught, and that without Jesus telling us what to teach, could not even exist. To try to raise a child without moral teaching and expect him to behave well was like raising him without Latin and expecting him to recite Virgil. Adam Smith begged to differ. He thought that morality owed little to teaching and nothing to reason, but evolved by a sort of reciprocal exchange within each person's mind as he or she grew from childhood, and within society. Morality therefore emerged as a consequence of certain aspects of human nature in response to social conditions.

As the Adam Smith scholar James Otteson has observed, Smith, who wrote a history of astronomy early in his career, saw himself as following explicitly in Newton's footsteps, both by looking for regularities in natural phenomena and by employing the parsimony principle of using as simple an explanation as possible. He praised Newton in his history of astronomy for the fact that he 'discovered that he could join together the movement of the planets by so familiar a principle of connection'. Smith was also part of a Scottish tradition that sought cause and effect in the history of a topic: instead of asking what is the perfect Platonic ideal of a moral system, ask rather how it came about.

It was exactly this modus operandi that Smith brought to moral philosophy. He wanted to understand where morality came from, and to explain it simply. As so often with Adam Smith, he deftly avoided the pitfalls into which later generations would fall. He saw straight through the nature-versus-nurture debate and came up with a nature-via-nurture explanation that was far ahead of its time. He starts *The Theory of Moral Sentiments* with a simple observation: we all enjoy making other people happy.

> How selfish soever man may be supposed, there are evidently some principles in his nature, which interest him in the fortunes of others, and render their happiness necessary to him, though he derives nothing from it, but the pleasure of seeing it.

And we all desire what he calls mutual sympathy of sentiments: 'Nothing pleases us more than to observe in other men a fellow-feeling with all the emotions of our own breast.' Yet the childless Smith observed that a child does not have a sense of morality, and has to find out the hard way that he or she is not the centre of the universe. Gradually, by trial and error, a child discovers what behaviour leads to mutual sympathy of sentiments, and therefore can make him or her happy by making others happy. It is through everybody accommodating their desires to those of others that a system of shared morality arises, according to Smith. An invisible hand (the phrase first appears in Smith's lectures on astronomy, then here in *Moral Sentiments* and once more in *The Wealth of Nations*) guides us towards a common moral code. Otteson explains that the hand is invisible, because people are not setting out to create a shared system of morality; they aim only to achieve mutual sympathy now with the people they are dealing with. The parallel with Smith's later explanation of the market is clear to see: both are phenomena that emerge from individual actions, but not from deliberate design.

Smith's most famous innovation in moral philosophy is the 'impartial spectator', who we imagine to be watching over us when we are required to be moral. In other words, just as we learn to be moral by judging others' reactions to our actions, so we can imagine those reactions by positing a neutral observer who embodies our conscience. What would a disinterested observer, who knows all the facts, think of our conduct? We get pleasure from doing what he recommends, and guilt from not doing so. Voltaire put it pithily: 'The safest course is to do nothing against one's conscience. With this secret, we can enjoy life and have no fear from death.'

How morality emerges

There is, note, no need for God in this philosophy. As a teacher of Natural Theology among other courses, Smith was no declared atheist, but occasionally he strays dangerously close to Lucretian scepticism. It is hardly surprising that he at least paid lip service to God, because three of his predecessors at Glasgow University, including Hutcheson, had been charged with heresy for not sticking to Calvinist orthodoxy. The mullahs of the day were vigilant. There remains one tantalising anecdote from a student, a disapproving John Ramsay, that Smith 'petitioned the Senatus . . . to be relieved of the duty of opening his class with a prayer', and, when refused, that his lectures led his students to 'draw an unwarranted conclusion, viz. that the great truths of theology, together with the duties which man owes to God and his neighbours, may be discovered in the light of nature without any special revelation'. The Adam Smith scholar Gavin Kennedy points out that in the sixth edition (1789) of *The Theory of Moral Sentiments*, published after his devout mother died, Smith excised or changed many religious references. He may have been a closet atheist, but he might also have been a theist, not taking Christianity literally, but assuming that some kind of god implanted benevolence in the human breast.

Morality, in Smith's view, is a spontaneous phenomenon, in the sense that people decide their own moral codes by seeking mutual sympathy of sentiments in society, and moralists then observe and record these conventions and teach them back to people as top–down instructions. Smith is essentially saying that the priest who tells you how to behave is basing his moral code on observations of what moral people actually do.

There is a good parallel with teachers of grammar, who do little more than codify the patterns they see in everyday speech and tell them back to us as rules. Only occasionally, as with split infinitives, do their rules go counter to what good writers do. Of course, it is possible for a priest to invent and promote a new rule of morality, just as it is possible for a language maven to invent and promote a new rule of grammar or syntax, but it is remarkably rare. In both cases, what happens is that usage changes and the teachers gradually go along with it, sometimes pretending to be the authors.

So, for example, in my lifetime, disapproval of homosexuality has become ever more morally unacceptable in the West, while disapproval of paedophilia has become ever more morally mandatory. Male celebrities who broke the rules with under-age girls long ago and thought little of it now find themselves in court and in disgrace; while others who broke the (then) rules with adult men long ago and risked disgrace can now openly speak of their love. Don't get me wrong: I approve of both these trends – but that's not my point. My point is that the changes did not come about because some moral leader or committee ordained them, at least not mainly, let alone that some biblical instruction to make the changes came to light. Rather, the moral negotiation among ordinary people gradually changed the common views in society, with moral teachers reflecting the changes along the way. Morality, quite literally, evolved. In just the same way, words like 'enormity' and 'prevaricate' have changed their meaning in my lifetime, though no committee met to consider an alteration in the meaning of the words, and there is very little the grammarians can do to prevent it. (Indeed, grammarians spend most of their time deploring linguistic innovation.) Otteson points out that Smith in his writing uses the word 'brothers' and 'brethren' interchangeably, with a slight preference for the latter. Today, however, the rules have changed, and you would only use 'brethren' for the plural of brothers if you were being affected, antiquarian or mocking.

Smith was acutely aware of this parallel with language, which is why he insisted on appending his short essay on the origin of language to his *Theory of Moral Sentiments* in its second and later editions. In the essay, Smith makes the point that the laws of language are an invention, rather than a discovery – unlike, say, the laws of physics. But they are still laws: children are corrected by their parents and their peers if they say 'bringed' instead of 'brought'. So language is an ordered system, albeit arrived at spontaneously through some kind of trial and error among people trying to make 'their mutual wants intelligible to each other'. Nobody is in charge, but the system is orderly. What a peculiar and novel idea. What a subversive thought. If God is not needed for morality, and if language is a spontaneous system, then perhaps the king, the pope and the official are not quite as vital to the functioning of an orderly society as they pretend?

As the American political scientist Larry Arnhart puts it, Smith is a founder of a key tenet of liberalism, because he rejects the Western tradition that morality must conform to a transcendental cosmic order, whether in the form of a cosmic God, a cosmic Reason, or a cosmic Nature. 'Instead of this transcendental moral cosmology, liberal morality is founded on an empirical moral anthropology, in which moral order arises from within human experience.'

Above all, Smith allows morality and language to change, to evolve. As Otteson puts it, for Smith, moral judgements are generalisations arrived at inductively on the basis of past experience. We log our own approvals and disapprovals of our own and others' conduct, and observe others doing the same. 'Frequently repeated patterns of judgement can come to have the appearance of moral duties or even commandments from on high, while patterns that recur with less frequency will enjoy commensurately less confidence.' It is in the messy empirical world of human experience that we find morality. Moral philosophers observe what we do; they do not invent it.

Better angels

Good grief. Here is an eighteenth-century, middle-class Scottish professor saying that morality is an accidental by-product of the way human beings adjust their behaviour towards each other as they grow up; saying that morality is an emergent phenomenon that arises spontaneously among human beings in a relatively peaceful society; saying that goodness does not need to be taught, let alone associated with the superstitious belief that it would not exist but for the divine origin of an ancient Palestinian carpenter. Smith sounds remarkably like Lucretius (whom he certainly read) in parts of his *Moral Sentiments* book, but he also sounds remarkably like Steven Pinker of Harvard University today discussing the evolution of society towards tolerance and away from violence.

As I will explore, there is in fact a fascinating convergence here. Pinker's account of morality growing strongly over time is, at bottom, very like Smith's. To put it at its baldest, a Smithian child, developing his sense of morality in a violent medieval society in Prussia (say) by trial and error, would end up with a moral code quite different from such a child growing up in a peaceful German (say) suburb today. The medieval person would be judged moral if he killed people in defence of his honour or his city; whereas today he would be thought moral if he refused meat and gave copiously to charity, and thought shockingly immoral if he killed somebody for any reason at all, and especially for honour. In Smith's evolutionary view of morality, it is easy to see how morality is relative and will evolve to a different end point in different societies, which is exactly what Pinker documents.

Pinker's book *The Better Angels of Our Nature* chronicles the astonishing and continuing decline in violence of recent centuries. We have just lived through the decade with the lowest global death rate in warfare on record; we have seen homicide rates fall by 99 per cent in most Western countries since medieval times; we have seen racial, sexual, domestic, corporal, capital and other forms of violence in headlong retreat; we have seen discrimination and prejudice go from normal to disgraceful; we have come to disapprove of all sorts of violence as entertainment, even against animals. This is not to say there is no violence left, but the declines that Pinker documents are quite remarkable, and our horror at the violence that still remains implies that the decline will continue. Our grandchildren will stand amazed at some of the things we still find quite normal.

To explain these trends, Pinker turns to a theory first elaborated by Norbert Elias, who had the misfortune to publish it as a Jewish refugee from Germany in Britain in 1939, shortly before he was interned by the British on the grounds that he was German. Not a good position from which to suggest that violence and coercion were diminishing. It was not until it was translated into English three decades later in 1969, in a happier time, that his theory was widely appreciated. Elias argued that a 'civilising process' had sharply altered the habits of Europeans since the Middle Ages, that as people became more urban, crowded, capitalist and secular, they became nicer too. He hit upon this paradoxical realisation – for which there is now, but was not then, strong statistical evidence – by combing the literature of medieval Europe and documenting the casual, frequent and routine violence that was then normal. Feuds flared into murders all the time; mutilation and death were common punishments; religion enforced its rules with torture and sadism; entertainments were often violent. Barbara Tuchman in her book *A Distant Mirror* gives an example of a popular game in medieval France: people with their hands tied behind their backs competed to kill a cat nailed to a post by battering it with their heads, risking the loss of an eye from the scratching of the desperate cat in the process. Ha ha.

Elias argued that moral standards evolved; to illustrate the point he documented the etiquette guides published by Erasmus and other philosophers. These guides are full of suggestions about table manners, toilet manners and bedside manners that seem unnecessary to state, but are therefore revealing: 'Don't greet someone while they are urinating or defecating . . . don't blow your nose on to the table-cloth or into your fingers, sleeve or hat . . . turn away when spitting lest your saliva fall on someone . . . don't pick your nose while eating.' In short, the very fact that these injunctions needed mentioning implies that medieval European life was pretty disgusting by modern standards. Pinker comments: 'These are the kind of directives you'd expect a parent to give to a three-year-old, not a great philosopher to a literate readership.' Elias argued that the habits of refinement, self-control and consideration that are second nature to us today had to be acquired. As time went by, people 'increasingly inhibited their impulses, anticipated the long-term consequences of their actions, and took other people's thoughts and feelings into consideration'. In other words, not blowing your nose on the tablecloth was all one with not stabbing your neighbour. It's a bit like a historical version of the broken-window theory: intolerance of small crimes leads to intolerance of big ones.

Doux commerce

But how were these gentler habits acquired? Elias realised that we have internalised the punishment for breaking these rules (and the ones against more serious violence) in the form of a sense of shame. That is to say, just as Adam Smith argued, we rely on an impartial spectator, and we learned earlier and earlier in life to see his point of view as he became ever more censorious. But why? Elias and Pinker give two chief reasons: government and commerce. With an increasingly centralised government focused on the king and his court, rather than local warlords, people had to behave more like courtiers and less like warriors. That meant not only less violent, but also more refined. Leviathan enforced the peace, if only to have more productive peasants to tax. Revenge for murder was nationalised as a crime to be punished, rather than privatised as a wrong to be righted. At the same time, commerce led people to value the opportunity to be trusted by a stranger in a transaction. With increasingly money-based interactions among strangers, people increasingly began to think of neighbours as potential trading partners rather than potential prey. Killing the shopkeeper makes no sense. So empathy, self-control and morality became second nature, though morality was always a double-edged sword, as likely to cause violence as to prevent it through most of history.

Lao Tzu saw this twenty-six centuries ago: 'The more prohibitions you have, the less virtuous people will be.' Montesquieu's phrase for the calming effect of trade on human violence, intolerance and enmity was '*doux commerce*' – sweet commerce. And he has been amply vindicated in the centuries since. The richer and more market-oriented societies have become, the nicer people have behaved. Think of the Dutch after 1600, the Swedes after 1800, the Japanese after 1945, the Germans likewise, the Chinese after 1978. The long peace of the nineteenth century coincided with the growth of free trade. The paroxysm of violence that convulsed the world in the first half of the twentieth century coincided with protectionism.

Countries where commerce thrives have far less violence than countries where it is suppressed. Does Syria suffer from a surfeit of commerce? Or Zimbabwe? Or Venezuela? Is Hong Kong largely peaceful because it eschews commerce? Or California? Or New Zealand? I once interviewed Pinker in front of an audience in London, and was very struck by the passion of his reply when an audience member insisted that profit was a form of violence and was on the increase. Pinker simply replied with a biographical story. His grandfather, born in Warsaw in 1900, emigrated to Montreal in 1926, worked for a shirt company (the family had made gloves in Poland), was laid off during the Great Depression, and then, with his grandmother, sewed neckties in his apartment, eventually earning enough to set up a small factory, which they ran until their deaths. And yes, it made a small profit (just enough to pay the rent and bring up Pinker's mother and her brothers), and no, his grandfather never hurt a fly. Commerce, he said, cannot be equated with violence.

'Participation in capitalist markets and bourgeois virtues has civilized the world,' writes Deirdre McCloskey in her book *The Bourgeois Virtues*. 'Richer and more urban people, contrary to what the magazines of opinion sometimes suggest, are *less* materialistic, *less* violent, *less* superficial than poor and rural people' (emphasis in original).

How is it then that conventional wisdom – especially among teachers and religious leaders – maintains that commerce is the cause of nastiness, not niceness? That the more we grow the economy and the more we take part in 'capitalism', the more selfish, individualistic and thoughtless we become? This view is so widespread it even leads such people to assume – against the evidence – that violence is on the increase. As Pope Francis put it in his 2013 apostolic exhortation *Evangelii Gaudium*, 'unbridled' capitalism has made the poor miserable even as it enriched the rich, and is responsible for the fact that 'lack of respect for others and violence are on the rise'. Well, this is just one of those conventional wisdoms that is plain wrong. There has been a decline in violence, not an increase, and it has been fastest in the countries with the least bridled versions of capitalism – not that there is such a thing as unbridled capitalism anywhere in the world. The ten most violent countries in the world in 2014 – Syria, Afghanistan, South Sudan, Iraq, Somalia, Sudan, Central African Republic, Democratic Republic of the Congo, Pakistan and North Korea – are all among the least capitalist. The ten most peaceful – Iceland, Denmark, Austria, New Zealand, Switzerland, Finland, Canada, Japan, Belgium and Norway – are all firmly capitalist.

My reason for describing Pinker's account of the Elias theory in such detail is because it is a thoroughly evolutionary argument. Even when Pinker credits Leviathan – government policy – for reducing violence, he implies that the policy is as much an attempt to reflect changing sensibility as to change sensibility. Besides, even Leviathan's role is unwitting: it did not set out to civilise, but to monopolise. It is an extension of Adam Smith's theory, uses Smith's historical reasoning, and posits that the moral sense, and the propensity to violence and sordid behaviour, evolve. They evolve not because somebody ordains that they should evolve, but spontaneously. The moral order emerges and continually changes. Of course, it can evolve towards greater violence, and has done so from time to time, but mostly it has evolved towards peace, as Pinker documents in exhaustive detail. In general, over the past five hundred years in Europe and much of the rest of the world, people became steadily less violent, more tolerant and more ethical, without even realising they were doing so. It was not until Elias spotted the trend in words, and later historians then confirmed it in statistics, that we even knew it was happening. It happened to us, not we to it.

The evolution of law

It is an extraordinary fact, unremembered by most, that in the Anglosphere people live by laws that did not originate with governments at all. British and American law derives ultimately from the common law, which is a code of ethics that was written by nobody and everybody. That is to say, unlike the Ten Commandments or most statute law, the common law emerges and evolves through precedent and adversarial argument. It 'evolves incrementally, rather than leaps convulsively or stagnates idly', in the words of legal scholar Allan Hutchinson. It is 'a perpetual work-in-progress – evanescent, dynamic, messy, productive, tantalizing, and bottom up'. The author Kevin Williamson reminds us to be astonished by this fact: 'The most successful, most practical, most cherished legal system in the world did not have an author. Nobody planned it, no sublime legal genius thought it up. It emerged in an iterative, evolutionary manner much like a language emerges.' Trying to replace the common law with a rationally designed law is, he jests, like trying to design a better rhinoceros in a laboratory.

Judges change the common law incrementally, adjusting legal doctrine case by case to fit the facts on the ground. When a new puzzle arises, different judges come to different conclusions about how to deal with it, and the result is a sort of genteel competition, as successive courts gradually choose which line they prefer. In this sense, the common law is built by natural selection.

Common law is a peculiarly English development, found mainly in countries that are former British colonies or have been influenced by the Anglo-Saxon tradition, such as Australia, India, Canada and the United States. It is a beautiful example of spontaneous order. Before the Norman Conquest, different rules and customs applied in different regions of England. But after 1066 judges created a common law by drawing on customs across the country, with an occasional nod towards the rulings of monarchs. Powerful Plantagenet kings such as Henry II set about standardising the laws to make them consistent across the country, and absorbed much of the common law into the royal courts. But they did not invent it. By contrast, European rulers drew on Roman law, and in particular a compilation of rules issued by the Emperor Justinian in the sixth century that was rediscovered in eleventh-century Italy. Civil law, as practised on the continent of Europe, is generally written by government.

In common law, the elements needed to prove the crime of murder, for instance, are contained in case law rather than defined by statute. To ensure consistency, courts abide by precedents set by higher courts examining the same issue. In civil-law systems, by contrast, codes and statutes are designed to cover all eventualities, and judges have a more limited role of applying the law to the case in hand. Past judgements are no more than loose guides. When it comes to court cases, judges in civil-law systems tend towards being investigators, while their peers in common-law systems act as arbiters between parties that present their arguments.

Which of these systems you prefer depends on your priorities. Jeremy Bentham argued that the common law lacked coherence and rationality, and was a repository of 'dead men's thoughts'. The libertarian economist Gordon Tullock, a founder of the public-choice school, argued that the common-law method of adjudication is inherently inferior because of its duplicative costs, inefficient means of ascertaining the facts, and scope for wealth-destroying judicial activism.

Others respond that the civil-law tradition, in its tolerance of arbitrary confiscation by the state and its tendency to mandate that which it does not outlaw, has proved less a friend of liberty than the common law. Friedrich Hayek advanced the view that the common law contributed to greater economic welfare because it was less interventionist, less under the tutelage of the state, and was better able to respond to change than civil legal systems; indeed, it was for him a legal system that led, like the market, to a spontaneous order.

A lot of Britain's continuing discomfort with the European Union derives from the contrast between the British tradition of bottom–up law-making and the top–down Continental version. The European Parliament member Daniel Hannan frequently reminds his colleagues of the bias towards liberty of the common law: 'This extraordinary, sublime idea that law does not emanate from the state but that rather there was a folk right of existing law that even the king and his ministers were subject to.'

The competition between these two traditions is healthy. But the point I wish to emphasise is that it is perfectly possible to have law that emerges, rather than is created. To most people that is a surprise. They vaguely assume in the backs of their minds that the law is always invented, rather than that it evolved. As the economist Don Boudreaux has argued, 'Law's expanse is so vast, its nuances so many and rich, and its edges so frequently changing that the popular myth that law is that set of rules designed and enforced by the state becomes increasingly absurd.'

It is not just the common law that evolves through replication, variation and selection. Even civil law, and constitutional interpretation, see gradual changes, some of which stick and some of which do not. The decisions as to which of these changes stick are not taken by omniscient judges, and nor are they random; they are chosen by the process of selection. As the legal scholar Oliver Goodenough argues, this places the evolutionary explanation at the heart of the system as opposed to appealing to an outside force. Both 'God made it happen' and 'Stuff happens' are external causes, whereas evolution is a 'rule-based cause internal to time and space as we experience them'.

3

The Evolution of Life

A mistake I strongly urge you to avoid for all you're worth,
An error in this matter you should give the widest berth:
Namely don't imagine that the bright lights of your eyes
Were purpose made so we could look ahead, or that our thighs
And calves were hinged together at the joints and set on feet
So we could walk with lengthy stride, or that forearms fit neat
To brawny upper arms, and are equipped on right and left
With helping hands, solely that we be dexterous and deft
At undertaking all the things we need to do to live,
This rationale and all the others like it people give,
Jumbles effect and cause, and puts the cart before the horse . . .

Lucretius, *De Rerum Natura*, Book 4, lines 823–33

Charles Darwin did not grow up in an intellectual vacuum. It is no accident that alongside his scientific apprenticeship he had a deep inculcation in the philosophy of the Enlightenment. Emergent ideas were all around him. He read his grandfather's Lucretius-emulating poems. 'My studies consist in Locke and Adam Smith,' he wrote from Cambridge, citing two of the most bottom–up philosophers. Probably it was Smith's *The Moral Sentiments* that he read, since it was more popular in universities than *The Wealth of Nations*. Indeed, one of the books that Darwin read in the autumn of 1838 after returning from the voyage of the *Beagle* and when about to crystallise the idea of natural selection was Dugald Stewart's biography of Adam Smith, from which he got the idea of competition and emergent order. The same month he read, or reread, the political economist Robert Malthus's essay on population, and was struck by the notion of a struggle for existence in which some thrived and others did not, an idea which helped trigger the insight of natural selection. He was friendly at the time with Harriet Martineau, a firebrand radical who campaigned for the abolition of slavery and also for the 'marvellous' free-market ideas of Adam Smith. She was a close confidante of Malthus. Through his mother's (and future wife's) family, the Wedgwoods, Darwin moved in a circle of radicalism, trade and religious dissent, meeting people like the free-market MP and thinker James Mackintosh. The evolutionary biologist Stephen Jay Gould once went so far as to argue that natural selection 'should be viewed as an extended analogy . . . to the laissez-faire economics of Adam Smith'. In both cases, Gould argued, balance and order emerged from the actions of individuals, not from external or divine control. As a Marxist, Gould surprisingly approved of this philosophy – for biology, but not for economics: 'It is ironic that Adam Smith's system of laissez faire does not work in his own domain of economics, for it leads to oligopoly and revolution.'

In short, Charles Darwin's ideas evolved, themselves, from ideas of emergent order in human society that were flourishing in early-nineteenth-century Britain. The general theory of evolution came before the special theory. All the same, Darwin faced a formidable obstacle in getting people to see undirected order in nature. That obstacle was the argument from design as set out, most ably, by William Paley.

In the last book that he published, in 1802, the theologian William Paley set out the argument for biological design based upon purpose. In one of the finest statements of design logic, from an indubitably fine mind, he imagined stubbing his toe against a rock while crossing a heath, then imagined his reaction if instead his toe had encountered a watch. Picking up the watch, he would conclude that it was man-made: 'There must have existed, at some time, and at some place or other, an artificer or artificers, who formed [the watch] for the purpose which we find it actually to answer; who comprehended its construction, and designed its use.' If a watch implies a watchmaker, then how could the exquisite purposefulness of an animal not imply an animal-maker? 'Every indication of contrivance, every manifestation of design, which existed in the watch, exists in the works of nature; with the difference, on the side of nature, of being greater or more, and that in a degree which exceeds all computation.'

Paley's argument from design was not new. It was Newton's logic applied to biology. Indeed, it was a version of one of the five arguments for the existence of God advanced by Thomas Aquinas six hundred years before: 'Whatever lacks intelligence cannot move towards an end, unless it be directed by some being endowed with knowledge and intelligence.' And in 1690 the high priest of common sense himself, John Locke, had effectively restated the same idea as if it were so rational that nobody could deny it. Locke found it 'as impossible to conceive that ever bare incogitative Matter should produce a thinking, intelligent being, as that nothing should produce Matter'. Mind came first, not matter. As Dan Dennett has pointed out, Locke gave an empirical, secular, almost mathematical stamp of approval to the idea that God was the designer.

Hume's swerve

The first person to dent this cosy consensus was David Hume. In a famous passage from his *Dialogues Concerning Natural Religion* (published posthumously in 1779), Hume has Cleanthes, his imaginary theist, state the argument from design in powerful and eloquent words:

> Look around the world: Contemplate the whole and every part of it: You will find it to be nothing but one great machine, subdivided into an infinite number of lesser machines . . . All these various machines, and even their most minute parts, are adjusted to each other with an accuracy, which ravishes into admiration all men, who have ever contemplated them. The curious adapting of means to ends, exceeds the productions of human contrivance; of human design, thought, wisdom, and intelligence. Since, therefore the effects resemble each other, we are led to infer, by all the rules of analogy, that the causes also resemble. [Dialogues, 2.5/143]

It's an inductive inference, Dennett points out: where there's design there's a designer, just as where there's smoke there's fire.

But Philo, Cleanthes's imaginary deist interlocutor, brilliantly rebuts the logic. First, it immediately prompts the question of who designed the designer. 'What satisfaction is there in that infinite progression?' Then he points out the circular reasoning: God's perfection explains the world's design, which proves God's perfection. And then, how do we know that God is perfect? Might he not have been a 'stupid mechanic, who imitated others' and 'botched and bungled' his way through different worlds during 'infinite ages of world making'? Or might not the same argument prove God to be multiple gods, or a 'perfect anthropomorphite' with human form, or an animal, or a tree, or a 'spider who spun the whole complicated mass from his bowels'?

Hume was now enjoying himself. Echoing the Epicureans, he began to pick holes in all the arguments of natural theology. A true believer, Philo said, would stress 'that there is a great and immeasurable, because incomprehensible, difference between the human and the divine mind', so it is idolatrous blasphemy to compare the deity to a mere engineer. An atheist, on the other hand, might be happy to concede the purposefulness of nature but explain it by some analogy other than a divine intelligence – as Charles Darwin eventually did.

In short, Hume, like Voltaire, had little time for divine design. By the time he finished, his alter ego Philo had effectively demolished the entire argument from design. Yet even Hume, surveying the wreckage, suddenly halted his assault and allowed the enemy forces to escape the field. In one of the great disappointments in all philosophy, Philo suddenly agrees with Cleanthes at the end, stating that if we are not content to call the supreme being God, then 'what can we call him but Mind or Thought'? It's Hume's Lucretian swerve. Or is it? Anthony Gottlieb argues that if you read it carefully, Hume has buried a subtle hint here, designed not to disturb the pious and censorious even after his death, that mind may be matter.

Dennett contends that Hume's failure of nerve cannot be explained by fear of persecution for atheism. He arranged to have his book published after his death. In the end it was sheer incredulity that caused him to balk at the ultimate materialist conclusion. Without the Darwinian insight, he just could not see a mechanism by which purpose came from matter.

Through the gap left by Hume stole William Paley. Philo had used the metaphor of the watch, arguing that pieces of metal could 'never arrange themselves so as to compose a watch'. Though well aware of Philo's objections, Paley still inferred a mind behind the watch on the heath. It was not that the watch was made of components, or that it was close to perfect in its design, or that it was incomprehensible – arguments that had appealed to a previous generation of physicists and that Hume had answered. It was that it was clearly designed to do a job, not individually and recently but once and originally in an ancestor. Switching metaphors, Paley asserted that 'there is precisely the same proof that the eye was made for vision, as there is that the telescope was made for assisting it'. The eyes of animals that live in water have a more curved surface than the eyes of animals that live on land, he pointed out, as befits the different refractive indices of the two elements: organs are adapted to the natural laws of the world, rather than vice versa.

But if God is omnipotent, why does he need to design eyes at all? Why not just give animals a magic power of vision without an organ? Paley had an answer of sorts. God could have done 'without the intervention of instruments or means: but it is in the construction of instruments, in the choice and adaptation of means, that a creative intelligence is seen'. God has been pleased to work within the laws of physics, so that we can have the pleasure of understanding them. In this way, Paley's modern apologists argue, God cannot be contradicted by the subsequent discovery of evolution by natural selection. He'd put that in place too to cheer us up by discovering it.

Paley's argument boils down to this: the more spontaneous mechanisms you discover to explain the world of living things, the more convinced you should be that there is an intelligence behind them. Confronted with such a logical contortion, I am reminded of one of the John Cleese characters in *Monty Python's Life of Brian*, when Brian denies that he is the Messiah: 'Only the true Messiah denies his divinity.'

Darwin on the eye

Nearly six decades after Paley's book, Charles Darwin's produced a comprehensive and devastating answer. Brick by brick, using insights from an Edinburgh education in bottom–up thinking, from a circumnavigation of the world collecting facts of stone and flesh, from a long period of meticulous observation and induction, he put together an astonishing theory: that the differential replication of competing creatures would produce cumulative complexity that fitted form to function without anybody ever comprehending the rationale in a mind. And thus was born one of the most corrosive concepts in all philosophy. Daniel Dennett in his book *Darwin's Dangerous Idea* compares

Darwinism to universal acid; it eats through every substance used to contain it. 'The creationists who oppose Darwinism so bitterly are right about one thing: Darwin's dangerous idea cuts much deeper into the fabric of our most fundamental beliefs than many of its sophisticated apologists have yet admitted, even to themselves.'

The beauty of Darwin's explanation is that natural selection has far more power than any designer could ever call upon. It cannot know the future, but it has unrivalled access to information about the past. In the words of the evolutionary psychologists Leda Cosmides and John Tooby, natural selection surveys 'the results of alternative designs operating in the real world, over millions of individuals, over thousands of generations, and weights alternatives by the statistical distribution of their consequences'. That makes it omniscient about what has worked in the recent past. It can overlook spurious and local results and avoid guesswork, inference or models: it is based on the statistical results of the actual lives of creatures in the actual range of environments they encounter.

One of the most perceptive summaries of Darwin's argument was made by one of his fiercest critics. A man named Robert Mackenzie Beverley, writing in 1867, produced what he thought was a devastating demolition of the idea of natural selection. Absolute ignorance is the artificer, he pointed out, trying to take the place of absolute wisdom in creating the world. Or (and here Beverley's fury drove him into capital letters), 'IN ORDER TO MAKE A PERFECT AND BEAUTIFUL MACHINE, IT IS NOT REQUISITE TO KNOW HOW TO MAKE IT.' To which Daniel Dennett, who is fond of this quotation, replies: yes, indeed! That is the essence of Darwin's idea: that beautiful and intricate organisms can be made without anybody knowing how to make them. A century later, an economist named Leonard Reed in an essay called 'I, Pencil', made the point that this is also true of technology. It is indeed the case that in order to make a perfect and beautiful machine, it is not requisite to know how to make it. Among the myriad people who contribute to the manufacture of a simple pencil, from graphite miners and lumberjacks to assembly-line workers and managers, not to mention those who grow the coffee that each of these drinks, there is not one person who knows how to make a pencil from scratch. The knowledge is held in the cloud, between brains, rather than in any individual head. This is one of the reasons, I shall argue in a later chapter, that technology evolves too.

Charles Darwin's dangerous idea was to take away the notion of intentional design from biology altogether and replace it with a mechanism that builds 'organized complexity . . . out of primeval simplicity' (in Richard Dawkins's words). Structure and function emerge bit by incremental bit and without resort to a goal of any kind. It's 'a process that was as patient as it was mindless' (Dennett). No creature ever set out mentally intending to see, yet the eye emerged as a means by which animals could see. There is indeed an adapted purposefulness in nature – it makes good sense to say that eyes have a function – but we simply lack the language to describe function that emerged from a backward-looking process, rather than a goal-directed, forward-looking, mind-first one. Eyes evolved, Darwin said, because in the past simple eyes that provided a bit of vision helped the survival and reproduction of their possessors, not because there was some intention on the part of somebody to achieve vision. All our functional phrases are top–down ones. The eye is 'for seeing', eyes are there 'so that' we can see, seeing is to eyes as typing is to keyboards. The language and its metaphors still imply skyhooks.

Darwin confessed that the evolution of the eye was indeed a hard problem. In 1860 he wrote to the American botanist Asa Gray: 'The eye to this day gives me a cold shudder, but when I think of the fine known gradation my reason tells me I ought to conquer the odd shudder.' In 1871 in his *Descent of Man*, he wrote: 'To suppose that the eye with all its inimitable contrivances for adjusting the focus to different distances, for admitting different amounts of light, and for the correction of spherical and chromatic aberration, could have been formed by natural selection, seems, I freely confess, absurd in the highest degree.'

But he then went on to set out how he justified the absurdity. First, the same could have been said of Copernicus. Common sense said the world stood still while the sun turned round it. Then he laid out how an eye could have emerged from nothing, step by step. He invoked 'numerous gradations' from a simple and imperfect eye to a complex one, 'each grade being useful to its possessor'. If such grades could be found among living animals, and they could, then there was no reason to reject natural selection, 'though insuperable by our imagination'. He had said something similar twenty-seven years before in his first, unpublished essay on natural selection: that the eye 'may possibly have been acquired by gradual selection of slight but in each case useful deviations'. To which his sceptical wife Emma had replied, in the margin: 'A great assumption'.

Pax optica

This is exactly what happened, we now know. Each grade was indeed useful to its possessor, because each grade still exists and still is useful to its owner. Each type of eye is just a slight improvement on the one before. A light-sensitive patch on the skin enables a limpet to tell which way is up; a light-sensitive cup enables a species called a slit-shelled mollusc to tell which direction the light is coming from; a pinhole chamber of light-sensitive cells enables the nautilus to focus a simple image of the world in good light; a simple lensed eye enables a murex snail to form an image even in low light; and an adjustable lens with an iris to control the aperture enables an octopus to perceive the world in glorious detail (the invention of the lens is easily explained, because any transparent tissue in the eye would have acted as partial refractor). Thus even just within the molluscs, every stage of the eye still exists, useful to each owner. How easy then to imagine each stage having existed in the ancestors of the octopus.

Richard Dawkins compares the progression through these grades to climbing a mountain (Mount Improbable) and at no point encountering a slope too steep to surmount. Mountains must be climbed from the bottom up. He shows that there are numerous such mountains – different kinds of eyes in different kinds of animal, from the compound eyes of insects to the multiple and peculiar eyes of spiders – each with a distinct range of partially developed stages showing how one can go step by step. Computer models confirm that there is nothing to suggest any of the stages would confer a disadvantage.

Moreover, the digitisation of biology since the discovery of DNA provides direct and unambiguous evidence of gradual evolution by the progressive alteration of the sequence of letters in genes. We now know that the very same gene, called Pax6, triggers the development of both the compound eye of insects and the simple eye of human beings. The two kinds of eye were inherited from a common ancestor. A version of a Pax gene also directs the development of simple eyes in jellyfish. The 'opsin' protein molecules that react to light in the eye can be traced back to the common ancestor of all animals except sponges. Around 700 million years ago, the gene for opsin was duplicated twice to give the three kinds of light-sensitive molecules we possess today. Thus every stage in the evolution of eyes, from the development of light-sensitive molecules to the emergence of lenses and colour vision, can be read directly from the language of the genes. Never has a hard problem in science been so comprehensively and emphatically solved as Darwin's eye dilemma. Shudder no more, Charles.

Astronomical improbability?

The evidence for gradual, undirected emergence of the opsin molecule by the stepwise alteration of the digital DNA language is strong. But there remains a mathematical objection. The opsin molecule is composed of hundreds of amino acids in a sequence specified by the appropriate gene. If one were to arrive at the appropriate sequence to give opsin its light-detecting properties by trial and error it would take either a very long time or a very large laboratory. Given that there are twenty types of amino acid, then a protein molecule with a hundred amino acids in its chain can exist in 10 to the power of 130 different sequences. That's a number far greater than the number of atoms in the universe, and far greater than the number of nanoseconds since the Big Bang. So it's just not possible for natural selection, however many organisms it has to play with for however long, to arrive at a design for an opsin molecule from scratch. And an opsin is just one of tens of thousands of proteins in the body.

Am I heading for a Lucretian swerve? Will I be forced to concede that the combinatorial vastness of the library of possible proteins makes it impossible for evolution to find ones that work? Far from it. We know that human innovation rarely designs things from scratch, but jumps from one technology to the 'adjacent possible' technology, recombining existing features. So it is taking small, incremental steps. And we know that the same is true of natural selection. So the mathematics is misleading. In a commonly used analogy, you are not assembling a Boeing 747 with a whirlwind in a scrapyard, you are adding one last rivet to an existing design. And here there has been a remarkable recent discovery that makes natural selection's task much easier.

In a laboratory in Zürich a few years ago, Andreas Wagner asked his student João Rodriguez to use a gigantic assembly of computers to work his way through a map of different metabolic networks to see how far he could get by changing just one step at a time. He chose the glucose system in a common gut bacterium, and his task was to change one link in the whole metabolic chain in such a way that it still worked – that the creature could still make sixty or so bodily ingredients from this one sugar. How far could he get? In species other than the gut bacterium there are thousands of different glucose pathways. How many of them are just a single step different from each other? Rodriguez found he got 80 per cent of the way through a library of a thousand different metabolic pathways at his first attempt, never having to change more than one step at a time and never producing a metabolic pathway that did not work. 'When João showed me the answer, my first reaction was disbelief,' wrote Wagner. 'Worried that this might be a fluke, I asked João for many more random walks, a thousand more, each preserving metabolic meaning, each leading as far as possible, each leaving in a different direction.' Same result.

Wagner and Rodriguez had stumbled upon a massive redundancy built into the biochemistry of bacteria – and people. Using the metaphor of a 'Library of Mendel', in which imaginary building are stored the unimaginably vast number of all possible genetic sequences, Wagner identified a surprising pattern. 'The metabolic library is packed to its rafters with books that tell the same story in different ways,' he writes. 'Myriad metabolic texts with the same meaning raise the odds of finding any one of them – myriad-fold. Even better, evolution does not just explore the metabolic library like a single casual browser. It crowdsources, employing huge populations of organisms that scour the library for new texts.' Organisms are crowds of readers going through the Library of Mendel to find texts that make sense.

Wagner points out that biological innovation must be both conservative and progressive, because as it redesigns the body, it cannot ever produce a non-functional organism. Turning microbes into mammals over millions of years is a bit like flying the Atlantic while rebuilding the plane to a new design. The globin molecule, for example, has roughly the same three-dimensional shape and roughly the same function in plants and insects, but the sequences of amino acids in the two are 90 per cent different.

Doubting Darwin still

Yet, despite this overwhelming evidence of emergence, the yearning for design still lures millions of people back into doubting Darwin. The American 'intelligent design' movement evolved directly from a fundamentalist drive to promote religion within schools, coupled with a devious 'end run' to circumvent the USA's constitutional separation between Church and state. It has largely focused upon the argument from design in order to try to establish that the complex functional arrangements of biology cannot be explained except by God. As Judge John Jones of Pennsylvania wrote in his judgement in the pivotal case of Kitzmiller vs Dover Area School District in 2005, although proponents of intelligent design 'occasionally suggest that the designer could be a space alien or a time-traveling cell biologist, no serious alternative to God as the designer has been proposed'. Tammy Kitzmiller was one of several Dover parents who objected to her child being taught 'intelligent design' on a par with Darwinism. The parents went to court, and got the school district's law overturned.

In the United States, fundamentalist Christians have challenged Darwinism in schools for more than 150 years. They pushed state legislatures into adopting laws that prohibited state schools from teaching evolution, a trend that culminated in the Scopes 'monkey trial' of 1925. The defendant, John Scopes, deliberately taught evolution illegally to bring attention to the state of Tennessee's anti-evolution law. Prosecuted by William Jennings Bryan and defended by Clarence Darrow, Scopes was found guilty and fined a paltry $100, and even that was overturned on a technicality at appeal. There is a widespread legend that Bryan's victory was pyrrhic, because it made him look ridiculous and Scopes's punishment was light. But this is a comforting myth told by saltwater liberals living on the coasts. In the American heartland, Scopes's conviction emboldened the critics of Darwin greatly. Far from being ridiculed into silence, the fundamentalists gained ground in the aftermath of the Scopes trial, and held that ground for decades within the educational system. Textbooks became very cautious about Darwinism.

It was not until 1968 that the United States Supreme Court struck down all laws preventing the teaching of evolution in schools. Fundamentalists then fell back on teaching 'creation science', a concoction of arguments that purported to find scientific evidence for biblical events such as Noah's flood. In 1987 the Supreme Court effectively proscribed the teaching of creationism on the grounds that it was religion, not science.

It was then that the movement reinvented itself as 'intelligent design', focusing on the old Aquinas–Paley argument from design in its simplest form. Creationists promptly rewrote their textbook *Of Pandas and People*, using an identical definition for intelligent design as had been used for creation science; and systematically replaced the words 'creationism' and 'creationist' with 'intelligent design' in 150 places. This went wrong in one case, resulting in a strange spelling mistake in the book, 'cdesign proponentsists', which came to be called the 'missing link' between the two movements. This 'astonishing' similarity between the two schools of thought was crucial in causing Judge John Jones to deem intelligent design religious rather than scientific as he struck down the Dover School District's law demanding equal time for intelligent design and evolution in 2005. Intelligent design, according to the textbook *Of Pandas and People*, argued that species came into existence abruptly, and through an intelligent agency, with their distinctive features already present: fish with fins and scales, birds with feathers.

Jones's long Opinion in 2005 was a definitive and conclusive demolition of a skyhook, all the more persuasive since it came from a Christian, Bush-appointed, politically conservative judge with no scientific training. Jones pointed out that the scientific revolution had rejected unnatural causes to explain natural phenomena, rejected appeal to authority, and rejected revelation, in favour of empirical evidence. He systematically took apart the evidence presented by Professor Michael Behe, the main scientific champion of intelligent design testifying for the defendants. Behe, in his book *Darwin's Black Box* and subsequent papers, had used two main arguments for the existence of an intelligent designer: irreducible complexity and the purposeful arrangement of parts. The flagellum of a bacterium, he argued, is driven by a molecular rotary motor of great complexity. Remove any part of that system and it will not work. The blood-clotting system of mammals likewise consists of a cascade of evolutionary events, none of which makes sense without the others. And the immune system was not only inexplicably complex, but a natural explanation was impossible.

It was trivial work for evolution's champions, such as Kenneth Miller, to dismantle these cases in the Dover trial to the satisfaction of the judge. A fully functional precursor of the bacterial flagellum with a different job, known as the Type III secretory system, exists in some organisms and could easily have been adapted to make a rotary motor while still partly retaining its original advantageous role. (In the same way, the middle-ear bones of mammals, now used for hearing, are direct descendants of bones that were once part of the jaw of early fish.) The blood-clotting cascade is missing one step in whales and dolphins, or three steps in puffer fish, and still works fine. And the immune system's mysterious complexity is yielding bit by bit to naturalistic explanations; what's left no more implicates an intelligent designer, or a time-travelling genetic engineer, than it does natural selection. At the trial Professor Behe was presented with fifty-eight peer-reviewed papers and nine books about the evolution of the immune system.

As for the purposeful arrangement of parts, Judge Jones did not mince his words: 'This inference to design based upon the appearance of a "purposeful arrangement of parts" is a completely subjective proposition, determined in the eye of each beholder and his/her viewpoint concerning the complexity of a system.' Which is really the last word on Newton, Paley, Behe, and for that matter Aquinas.

More than 2,000 years ago Epicureans like Lucretius seem to have cottoned on to the power of natural selection, an idea that they probably got from the flamboyant Sicilian philosopher Empedocles (whose verse style was also a model for Lucretius), born in around 490 BC. Empedocles talked of animals that survived 'being organised spontaneously in a fitting way; whereas those which grew otherwise perished and continue to perish'. It was, had Empedocles only known it, probably the best idea he ever had, though he never seems to have followed it through. Darwin was rediscovering an idea.

Gould's swerve

Why was it even necessary, nearly 150 years after Darwin set out his theory, for Judge Jones to make the case again? This remarkable persistence of resistance to the idea of evolution, packaged and repackaged as natural theology, then creation science, then intelligent design, has never been satisfactorily explained. Biblical literalism cannot fully justify why people so dislike the idea of spontaneous biological complexity. After all, Muslims have no truck with the idea that the earth is 6,000 years old, yet they too find the argument from design persuasive. Probably fewer than 20 per cent of people in most Muslim-dominated countries accept Darwinian evolution to be true. Adnan Oktar, for example, a polemical Turkish creationist who also uses the name Harun Yahya, employs the argument from design to 'prove' that Allah created living things. Defining design as 'a harmonious assembling of various parts into an orderly form towards a common goal', he then argues that birds show evidence of design, their hollowed bones, strong muscles and feathers making it 'obvious that the bird is product of a certain design'. Such a fit between form and function, however, is very much part of the Darwinian argument too.

Secular people, too, often jib at accepting the idea that complex organs and bodies can emerge without a plan. In the late

1970s a debate within Darwinism, between a predominantly American school led by the fossil expert Stephen Jay Gould and a predominantly British one led by the behaviour expert Richard Dawkins, about the pervasiveness of adaptation, led to some bitter and high-profile exchanges. Dawkins thought that almost every feature of a modern organism had probably been subject to selection for a function, whereas Gould thought that lots of change happened for accidental reasons. By the end, Gould seemed to have persuaded many lay people that Darwinism had gone too far; that it was claiming a fit between form and function too often and too glibly; that the idea of the organism adapting to its environment through natural selection had been refuted or at least diminished. In the media, this fed what John Maynard Smith called 'a strong wish to believe that the Darwinian theory is false', and culminated in an editorial in the *Guardian* announcing the death of Darwinism.

Within evolutionary biology, however, Gould lost the argument. Asking what an organ had evolved to do continued to be the main means by which biologists interpreted anatomy, biochemistry and behaviour. Dinosaurs may have been large 'to' achieve stable temperatures and escape predation, while nightingales may sing 'to' attract females.

This is not the place to retell the story of that debate, which had many twists and turns, from the spandrels of the Cathedral of San Marco in Venice to the partial resemblance of a caterpillar to a bird's turd. My purpose here is different – to discern the motivation of Gould's attack on adaptationism and its extraordinary popularity outside science. It was Gould's Lucretian swerve. Daniel Dennett, Darwin's foremost philosopher, thought Gould was 'following in a long tradition of eminent thinkers who have been seeking skyhooks – and coming up with cranes', and saw his antipathy to 'Darwin's dangerous idea as fundamentally a desire to protect or restore the Mind-first, top–down vision of John Locke'.

Whether this interpretation is fair or not, the problem Darwin and his followers have is that the world is replete with examples of deliberate design, from watches to governments. Some of them even involve design: the many different breeds of pigeons that Darwin so admired, from tumblers to fantails, were all produced by 'mind-first' selective breeding, just like natural selection but at least semi-deliberate and intentional. Darwin's reliance on pigeon breeding to tell the tale of natural selection was fraught with danger – for his analogy was indeed a form of intelligent design.

Wallace's swerve

Again and again, Darwin's followers would go only so far, before swerving. Alfred Russel Wallace, for instance, co-discovered natural selection and was in many ways an even more radical enthusiast for Darwinism (a word he coined) than Darwin himself. Wallace was not afraid to include human beings within natural selection very early on; and he was almost alone in defending natural selection as the main mechanism of evolution in the 1880s, when it was sharply out of fashion. But then he executed a Lucretian swerve. Saying that 'the Brain of the Savage [had been] shown to be Larger than he Needs it to be' for survival, he concluded that 'a superior intelligence has guided the development of man in a definite direction, and for a special purpose'. To which Darwin replied, chidingly, in a letter: 'I hope you have not murdered too completely your own and my child.'

Later, in a book published in 1889 that resolutely champions Darwinism (the title of the book), Wallace ends by executing a sudden U-turn, just like Hume and so many others. Having demolished skyhook after skyhook, he suddenly erects three at the close. The origin of life, he declares, is impossible to explain without a mysterious force. It is 'altogether preposterous' to argue that consciousness in animals could be an emergent consequence of complexity. And mankind's 'most characteristic and noble faculties could not possibly have been developed by means of the same laws which have determined the progressive development of the organic world in general'. Wallace, who was by now a fervent spiritualist, demanded three skyhooks to explain life, consciousness and human mental achievements. These three stages of progress pointed, he said, to an unseen universe, 'a world of spirit, to which the world of matter is altogether subordinate'.

The lure of Lamarck

The repeated revival of Lamarckian ideas to this day likewise speaks of a yearning to reintroduce mind-first intentionality into Darwinism. Jean-Baptiste de Lamarck suggested long before Darwin that creatures might inherit acquired characteristics – so a blacksmith's son would inherit his father's powerful forearms even though these were acquired by exercise, not inheritance. Yet people obviously do not inherit mutilations from their parents, such as amputated limbs, so for Lamarck to be right there would have to be some kind of intelligence inside the body deciding what was worth passing on and what was not. But you can see the appeal of such a scheme to those left disoriented by the departure of God the designer from the Darwinised scene. Towards the end of his life, even Darwin flirted with some tenets of Lamarckism as he struggled to understand heredity.

At the end of the nineteenth century, the German biologist August Weismann pointed out a huge problem with Lamarckism: the separation of germ-line cells (the ones that end up being eggs or sperm) from other body cells early in the life of an animal makes it virtually impossible for information to feed back from what happens to a body during its life into its very recipe. Since the germ cells were not an organism in microcosm, the message telling them to adopt an acquired character must, Weismann argued, be of an entirely different nature from the change itself. Changing a cake after it has been baked cannot alter the recipe that was used.

The Lamarckians did not give up, though. In the 1920s a

herpetologist named Paul Kammerer in Vienna claimed to have changed the biology of midwife toads by changing their environment. The evidence was flaky at best, and wishfully interpreted. When accused of fraud, Kammerer killed himself. A posthumous attempt by the writer Arthur Koestler to make Kammerer into a martyr to the truth only reinforced the desperation so many non-scientists felt to rescue a top–down explanation of evolution.

It is still going on. Epigenetics is a respectable branch of genetic science that examines how modifications to DNA sequences acquired early in life in response to experience can affect the adult body. There is a much more speculative version of the story, though. Most of these modifications are swept clean when the sperm and egg cells are made, but perhaps a few just might survive the jump into a new generation. Certain genetic disorders, for example, seem to manifest themselves differently according to whether the mutant chromosome was inherited from the mother or the father – implying a sex-specific 'imprint' on the gene. And one study seemed to find a sex-specific effect on the mortality of Swedes according to how hungry their grandparents were when young. From a small number of such cases, none with very powerful results, some modern Lamarckians began to make extravagant claims for the vindication of the eighteenth-century French aristocrat. 'Darwinian evolution can include Lamarckian processes,' wrote Eva Jablonka and Marion Lamb in 2005, 'because the heritable variation on which selection acts is not entirely blind to function; some of it is induced or "acquired" in response to the conditions of life.'

But the evidence for these claims remains weak. All the data suggest that the epigenetic state of DNA is reset in each generation, and that even if this fails to happen, the amount of information imparted by epigenetic modifications is a minuscule fraction of the information imparted by genetic information. Besides, ingenious experiments with mice show that all the information required to reset the epigenetic modifications themselves actually lies in the genetic sequence. So the epigenetic mechanisms must themselves have evolved by good old Darwinian random mutation and selection. In effect, there is no escape to intentionality to be found here. Yet the motive behind the longing to believe in epigenetic Lamarckism is clear. As David Haig of Harvard puts it, 'Jablonka and Lamb's frustration with neo-Darwinism is with the pre-eminence that is ascribed to undirected, random sources of heritable variation.' He says he is 'yet to hear a coherent explanation of how the inheritance of acquired characters can, by itself, be a source of intentionality'. In other words, even if you could prove some Lamarckism in epigenetics, it would not remove the randomness.

Culture-driven genetic evolution

In fact, there is a way for acquired characteristics to come to be incorporated into genetic inheritance, but it takes many generations and it is blindly Darwinian. It goes by the name of the Baldwin effect. A species that over many generations repeatedly exposes itself to some experience will eventually find its offspring selected for a genetic predisposition to cope with that experience. Why? Because the offspring that by chance happen to start with a predisposition to cope with that circumstance will survive better than others. The genes can thereby come to embody the experience of the past. Something that was once learned can become an instinct.

A similar though not identical phenomenon is illustrated by the ability to digest lactose sugar in milk, which many people with ancestors from western Europe and eastern Africa possess. Few adult mammals can digest lactose, since milk is not generally drunk after infancy. In two parts of the world, however, human beings evolved the capacity to retain lactose digestion into adulthood by not switching off genes for lactase enzymes. These happened to be the two regions where the domestication of cattle for milk production was first invented. What a happy coincidence! Because people could digest lactose, they were able to invent dairy farming? Well no, the genetic switch plainly happened as a consequence, not a cause, of the invention of dairy farming. But it still had to happen through random mutation followed by non-random survival. Those born by chance with the mutation that caused persistence of lactose digestion tended to be stronger and healthier than their siblings and rivals who could digest less of the goodness in milk. So they thrived, and the gene for lactose digestion spread rapidly. On closer inspection, this incorporation of ancestral experience into the genes is all crane and no skyhook.

So incredible is the complexity of the living world, so counter-intuitive is the idea of boot-strapped, spontaneous intricacy, that even the most convinced Darwinian must, in the lonely hours of the night, have moments of doubt. Like Screwtape the devil whispering in the ear of a believer, the 'argument from personal incredulity' (as Richard Dawkins calls it) can be very tempting, even if you remind yourself that it's a massive non sequitur to find divinity in ignorance.

4

The Evolution of Genes

For certainly the elements of things do not collect
And order their formations by their cunning intellect,
Nor are their motions something they agree upon or propose;
But being myriad and many-mingled, plagued by blows
And buffeted through the universe for all time past,
By trying every motion and combination, they at last,
Fell into the present form in which the universe appears.

Lucretius, *De Rerum Natura*, Book 1, lines 1021–7

An especially seductive chunk of current ignorance is that concerning the origin of life. For all the confidence with which biologists trace the emergence of complex organs and organisms from simple proto-cells, the first emergence of those proto-cells is still shrouded in darkness. And where people are baffled, they are often tempted to resort to mystical explanations. When the molecular biologist Francis Crick, that most materialist of scientists, started speculating about 'panspermia' in the 1970s – the idea that life perhaps originated elsewhere in the universe and got here by microbial seeding – many feared that he was turning a little mystical. In fact he was just making an argument about probability: that it was highly likely, given the youth of the earth relative to the age of the universe, that some other planet got life before us and infected other solar systems. Still, he was emphasising the impenetrability of the problem.

Life consists of the capacity to reverse the drift towards entropy and disorder, at least locally – to use *information* to make local *order* from chaos while expending *energy*. Essential to these three skills are three kinds of molecule in particular – DNA for storing information, protein for making order, and ATP as the medium of energy exchange. How these came together is a chicken-and-egg problem. DNA cannot be made without proteins, nor proteins without DNA. As for energy, a bacterium uses up fifty times its own body weight in ATP molecules in each generation. Early life must have been even more profligate, yet would have had none of the modern molecular machinery for harnessing and storing energy. Wherever did it find enough ATP?

The crane that seems to have put these three in place was probably RNA, a molecule that still plays many key roles in the cell, and that can both store information like DNA, and catalyse reactions like proteins do. Moreover, RNA is made of units of base, phosphate and ribose sugar, just as ATP is. So the prevailing theory holds that there was once an 'RNA World', in which living things had RNA bodies with RNA genes, using RNA ingredients as an energy currency. The problem is that even this system is so fiendishly complex and interdependent that it's hard to imagine it coming into being from scratch. How, for example, would it have avoided dissipation: kept together its ingredients and concentrated its energy without the boundaries provided by a cell membrane? In the 'warm little pond' that Charles Darwin envisaged for the beginning of life, life would have dissolved away all too easily.

Don't give up. Until recently the origin of the RNA World seemed so difficult a problem that it gave hope to mystics; John Horgan wrote an article in *Scientific American* in 2011 entitled 'Psst! Don't Tell the Creationists, But Scientists Don't Have a Clue How Life Began'.

Yet today, just a few years later, there's the glimmer of a solution. DNA sequences show that at the very root of life's family tree are simple cells that do not burn carbohydrates like the rest of us, but effectively charge their electrochemical batteries by converting carbon dioxide into methane or the organic compound acetate. If you want to find a chemical environment that echoes the one these chemi-osmotic microbes have within their cells, look no further than the bottom of the Atlantic Ocean. In the year 2000, explorers found hydrothermal vents on the mid-Atlantic ridge that were quite unlike those they knew from other geothermal spots on the ocean floor. Instead of very hot, acidic fluids, as are found at 'black-smoker' vents, the new vents – known as the Lost City Hydrothermal Field – are only warm, are highly alkaline, and appear to last for tens of thousands of years. Two scientists, Nick Lane and William Martin, have begun to list the similarities between these vents and the insides of chemi-osmotic cells, finding uncanny echoes of life's method of storing energy. Basically, cells store energy by pumping electrically charged particles, usually sodium or hydrogen ions, across membranes, effectively creating an electrical voltage. This is a ubiquitous and peculiar feature of living creatures, but it appears it might have been borrowed from vents like those at Lost City.

Four billion years ago the ocean was acidic, saturated with carbon dioxide. Where the alkaline fluid from the vents met the acidic water, there was a steep proton gradient across the thin iron-nickel-sulphur walls of the pores that formed at the vents. That gradient had a voltage very similar in magnitude to the one in a modern cell. Inside those mineral pores, chemicals would have been trapped in a space with abundant energy, which could have been used to build more complex molecules. These in turn – as they began to accidentally replicate themselves using the energy from the proton gradients – became gradually more susceptible to a pattern of survival of the fittest. And the rest, as Daniel Dennett would say, is algorithm. In short, an emergent account of the origin of life is almost within reach.

All crane and no skyhook

As I mentioned earlier, the diagnostic feature of life is that it captures energy to create order. This is also a hallmark of civilisation. Just as each person uses energy to make buildings and devices and ideas, so each gene uses energy to make a structure of protein. A bacterium is limited in how large it can grow by the quantity of energy available to each gene. That's because the energy is captured at the cell membrane by pumping protons across the membrane, and the bigger the cell, the smaller its surface area relative to its volume. The only bacteria that grow big enough to be seen by the naked eye are ones that have huge empty vacuoles inside them.

However, at some point around two billion years after life started, huge cells began to appear with complicated internal structures; we call them eukaryotes, and we (animals as well as plants, fungi and protozoa) are them.

Nick Lane argues that the eukaryotic (r)evolution was made possible by a merger: a bunch of bacteria began to live inside an archeal cell (a different kind of microbe). Today the descendants of these bacteria are known as mitochondria, and they generate the energy we need to live. During every second of your life your thousand trillion mitochondria pump a billion trillion protons across their membranes, capturing the electrical energy needed to forge your proteins, DNA and other macromolecules.

Mitochondria still have their own genes, but only a small number – thirteen in us. This simplification of their genome was vital. It enabled them to generate far more surplus energy to support the work of 'our genome', which is what enables us to have complex cells, complex tissues and complex bodies. As a result we eukaryotes have tens of thousands of times more energy available per gene, making each of our genes capable of far greater productivity. That allows us to have larger cells as well as more complex structures. In effect, we overcame the size limit of the bacterial cell by hosting multiple internal membranes in mitochondria, and then simplifying the genomes needed to support those membranes.

There is an uncanny echo of this in the Industrial (R)evolution. In agrarian societies, a family could grow just enough food to feed itself, but there was little left over to support anybody else. So only very few people could have castles, or velvet coats, or suits of armour, or whatever else needed making with surplus energy. The harnessing of oxen, horses, wind and water helped generate a bit more surplus energy, but not much. Wood was no use – it provided heat, not work. So there was a permanent limit on how much a society could make in the way of capital – structures and things.

Then in the Industrial (R)evolution an almost inexhaustible supply of energy was harnessed in the form of coal. Coal miners, unlike peasant farmers, produced vastly more energy than they consumed. And the more they dug out, the better they got at it. With the first steam engines, the barrier between heat and work was breached, so that coal's energy could now amplify the work of people. Suddenly, just as the eukaryotic (r)evolution vastly increased the amount of energy per gene, so the Industrial (R)evolution vastly increased the amount of energy per worker. And that surplus energy, so the energy economist John Constable argues, is what built (and still builds) the houses, machines, software and gadgets – the capital – with which we enrich our lives. Surplus energy is indispensable to modern society, and is the symptom of wealth. An American consumes about ten times as much energy as a Nigerian, which is the same as saying he is ten times richer. 'With coal almost any feat is possible or easy,' wrote William Stanley Jevons; 'without it we are thrown back into the laborious poverty of early times.' Both the evolution of surplus energy generation by eukaryotes, and the evolution of surplus energy by industrialisation, were emergent, unplanned phenomena.

But I digress. Back to genomes. A genome is a digital computer program of immense complexity. The slightest mistake would alter the pattern, dose or sequence of expression of its 20,000 genes (in human beings), or affect the interaction of its hundreds of thousands of control sequences that switch genes on and off, and result in disastrous deformity or a collapse into illness. In most of us, for an incredible eight or nine decades, the computer program runs smoothly with barely a glitch.

Consider what must happen every second in your body to keep the show on the road. You have maybe ten trillion cells, not counting the bacteria that make up a large part of your body. Each of those cells is at any one time transcribing several thousand genes, a procedure that involves several hundred proteins coming together in a specific way and catalysing tens of chemical reactions for each of millions of base pairs. Each of those transcripts generates a protein molecule, thousands of amino acids long, which it does by entering a ribosome, a machine with tens of moving parts, capable of catalysing a flurry of chemical reactions. The proteins themselves then fan out within and without cells to speed reactions, transport goods, transmit signals and prop up structures. Millions of trillions of these immensely complicated events are occurring every second in your body to keep you alive, very few of which go wrong. It's like the world economy in miniature, only even more complex.

It is hard to shake the illusion that for such a computer to run such a program, there must be a programmer. Geneticists in the early days of the Human Genome Project would talk of 'master genes' that commanded subordinate sequences. Yet no such master gene exists, let alone an intelligent programmer. The entire thing not only emerged piece by piece through evolution, but runs in a democratic fashion too. Each gene plays its little role; no gene comprehends the whole plan. Yet from this multitude of precise interactions results a spontaneous design of unmatched complexity and order. There was never a better illustration of the validity of the Enlightenment dream – that order can emerge where nobody is in charge. The genome, now sequenced, stands as emphatic evidence that there can be order and complexity without any management.

On whose behalf?

Let's assume for the sake of argument that I have persuaded you that evolution is not directed from above, but is a self-organising process that produces what Daniel Dennett calls 'free-floating rationales' for things. That is to say, for example, a baby cuckoo pushes the eggs of its host from the nest in order that it can monopolise its foster parents' efforts to feed it, but nowhere has that rationale ever existed as a thought either in the mind of the cuckoo or in the mind of a cuckoo's designer. It exists now in your mind and mine, but only after the fact. Bodies and behaviours teem with apparently purposeful function that was never foreseen or planned. You will surely agree that this model can apply within the human genome, too; your blood-clotting genes are there to make blood-clotting proteins, the better to clot blood at a wound; but that functional design does not imply an intelligent designer who foresaw the need for blood clotting.

I'm now going to tell you that you have not gone far enough. God is not the only skyhook. Even the most atheistic scientist, confronted with facts about the genome, is tempted into command-and-control thinking. Here's one, right away: the idea that genes are recipes patiently waiting to be used by the cook that is the body. The collective needs of the whole organism are what the genes are there to serve, and they are willing slaves. You find this assumption behind almost any description of genetics – including ones by me – yet it is misleading. For it is just as truthful to turn the image upside down. The body is the plaything and battleground of genes at least as much as it is their purpose. Whenever somebody asks what a certain gene is for, they automatically assume that the question relates to the needs of the body: what is it for, in terms of the body's needs? But there are plenty of times when the answer to that question is 'The gene itself.'

The scientist who first saw this is Richard Dawkins. Long

before he became well known for his atheism, Dawkins was famous for the ideas set out in his book *The Selfish Gene*. 'We are survival machines – robot vehicles blindly programmed to preserve the selfish molecules known as genes,' he wrote. 'This is a truth that still fills me with astonishment.' He was saying that the only way to understand organisms was to see them as mortal and temporary vehicles used to perpetuate effectively immortal digital sequences written in DNA. A male deer risks its life in a battle with another stag, or a female deer exhausts her reserves of calcium producing milk for her young, not to help its own body's survival but to pass the genes to the next generation. Far from preaching selfish behaviour, therefore, this theory explains why we are often altruistic: it is the selfishness of the genes that enables individuals to be selfless. A bee suicidally stinging an animal that threatens the hive is dying for its country (or hive) so that its genes may survive – only in this case the genes are passed on indirectly, through the stinger's mother, the queen. It makes more sense to see the body as serving the needs of the genes than vice versa. Bottom–up.

One paragraph of Dawkins's book, little noticed at the time, deserves special attention. It has proved to be the founding text of an extremely important theory. He wrote:

Sex is not the only apparent paradox that becomes less puzzling the moment we learn to think in selfish gene terms. For instance, it appears that the amount of DNA in organisms is more than is strictly necessary for building them: a large fraction of the DNA is never translated into protein. From the point of view of the individual this seems paradoxical. If the 'purpose' of DNA is to supervise the building of bodies it is surprising to find a large quantity of DNA which does no such thing. Biologists are racking their brains trying to think what useful task this apparently surplus DNA is doing. From the point of view of the selfish genes themselves, there is no paradox. The true 'purpose' of DNA is to survive, no more and no less. The simplest way to explain the surplus DNA is to suppose that it is a parasite, or at best a harmless but useless passenger, hitching a ride in the survival machines created by the other DNA.

One of the people who read that paragraph and began thinking about it was Leslie Orgel, a chemist at the Salk Institute in California. He mentioned it to Francis Crick, who mentioned it in an article about the new and surprising discovery of 'split genes' – the fact that most animal and plant genes contain long sequences of DNA called 'introns' that are discarded after transcription. Crick and Orgel then wrote a paper expanding on Dawkins's selfish DNA explanation for all the extra DNA. So, at the same time, did the Canadian molecular biologists Ford Doolittle and Carmen Sapienza. 'Sequences whose only "function" is self-preservation will inevitably arise and be maintained,' wrote the latter. The two papers were published simultaneously in 1980.

It turns out that Dawkins was right. What would his theory predict? That the spare DNA would have features that made it good at getting itself duplicated and re-inserted into chromosomes. Bingo. The commonest gene in the human genome is the recipe for reverse transcriptase, an enzyme that the human body has little or no need for, and whose main function is usually to help the spread of retroviruses. Yet there are more copies and half-copies of this gene than of all other human genes combined. Why? Because reverse transcriptase is a key part of any DNA sequence that can copy itself and distribute the copies around the genome. It's a sign of a digital parasite. Most of the copies are inert these days, and some are even put to good use, helping to regulate real genes or bind proteins. But they are there because they are good at being there.

The skyhook here is a sort of cousin of Locke's 'mind-first' thinking: the assumption that the human good is the only good pursued within our bodies. The alternative view, penetratingly articulated by Dawkins, takes the perspective of the gene itself:

how DNA would behave if it could. Close to half of the human genome consists of so-called transposable elements designed to use reverse transcriptase. Some of the commonest are known by names like LINEs (17 per cent of the genome), SINEs (11 per cent) and LTR retrotransposons (8 per cent). Actual genes, by contrast, fill just 2 per cent of the genome. These transposons are sequences that are good at getting themselves copied, and there is no longer a smidgen of doubt that they are (mostly inert) digital parasites. They are not there for the needs of the body at all.

Junk is not the same as garbage

There is a close homology with computer viruses, which did not yet exist when Dawkins suggested the genetic version of the concept of digital parasitism. Some of the transposons, the SINEs, appear to be parasites of parasites, because they use the apparatus of longer, more complete sequences to get themselves disseminated. For all the heroic attempts to see their function in terms of providing variability that might one day lead to a brave new mutation, the truth is that their more immediate and frequent effect is occasionally to disrupt the reading of genes.

Of course, these selfish DNA sequences can thrive only because a small percentage of the genome does something much more constructive – builds a body that grows, learns and adapts sufficiently to its physical and social environment that it can eventually thrive, attract a mate and have babies. At which point the selfish DNA says, 'Thank you very much, we'll be making up half the sequence in the children too.'

It is currently impossible to explain the huge proportion of the human genome devoted to these transposons except by reference to the selfish DNA theory. There's just no other theory that comes close to fitting the facts. Yet it is routinely rejected, vilified and 'buried' by commentators on the fringe of genetics. The phrase that really gets their goat is 'junk DNA'. It's almost impossible to read an article on the topic without coming across surprisingly passionate denunciations of the 'discredited' notion that some of the DNA in a genome is useless. 'We have long felt that the current disrespectful (in a vernacular sense) terminology of junk DNA and pseudogenes,' wrote Jürgen Brosius and Stephen Jay Gould in an early salvo in 1992, 'has been masking the central evolutionary concept that features of no current utility may hold crucial evolutionary importance as recruitable sources of future change.' Whenever I write about this topic, I am deluged with moralistic denunciations of the 'arrogance' of scientists for rejecting unknown functions of DNA sequences. To which I reply: functions for whom? The body or the sequences?

This moral tone to the disapproval of 'so-called' junk DNA is common. People seem to be genuinely offended by the phrase. They sound awfully like the defenders of faith confronted with evolution – it's the bottom–up nature of the story that they dislike. Yet as I shall show, selfish DNA and junk DNA are both about as accurate as metaphors ever can be. And junk is not the same as garbage.

What's the fuss about? In the 1960s, as I mentioned earlier, molecular biologists began to notice that there seemed to be far more DNA in a cell than was necessary to make all the proteins in the cell. Even with what turned out to be a vastly over-inflated estimate of the number of genes in the human genome – then thought to be more than 100,000, now known to be about 20,000 – genes and their control sequences could account for only a small percentage of the total weight of DNA present in a cell's chromosomes, at least in mammals. It's less than 3 per cent in people. Worse, there was emerging evidence that we human beings did not seem to have the heaviest genomes or the most DNA. Humble protozoa, onions and salamanders have far bigger genomes. Grasshoppers have three times as much; lung-fish forty times as much. Known by the obscure name of the 'c-value paradox', this enigma exercised the minds of some of the most eminent scientists of the day. One of them, Susumu

Ohno, coined the term 'junk DNA', arguing that much of the DNA might not be under selection – that is to say, might not be being continuously honed by evolution to fit a function of the body.

He was not saying it was garbage. As Sydney Brenner later made plain, people everywhere make the distinction between two kinds of rubbish: 'garbage' which has no use and must be disposed of lest it rot and stink, and 'junk', which has no immediate use but does no harm and is kept in the attic in case it might one day be put back to use. You put garbage in the rubbish bin; you keep junk in the attic or garage.

Yet the resistance to the idea of junk DNA mounted. As the number of human genes steadily shrank in the 1990s and 2000s, so the desperation to prove that the rest of the genome must have a use (for the organism) grew. The new simplicity of the human genome bothered those who liked to think of the human being as the most complex creature on the planet. Junk DNA was a concept that had to be challenged. The discovery of RNA-coding genes, and of multiple control sequences for adjusting the activity of genes, seemed to offer some straws of hope to grasp. When it became clear that on top of the 5 per cent of the genome that seemed to be specifically protected from change between human beings and related species, another 4 per cent showed some evidence of being under selection, the prestigious journal *Science* was moved to proclaim 'no more junk DNA'. What about the other 91 per cent?

In 2012 the anti-junk campaign culminated in a raft of hefty papers from a huge consortium of scientists called ENCODE. These were greeted, as intended, with hype in the media announcing the Death of Junk DNA. By defining non-junk as any DNA that had something biochemical happen to it during normal life, they were able to assert that about 80 per cent of the genome was functional. (And this was in cancer cells, with abnormal patterns of DNA hyperactivity.) That still left 20 per cent with nothing going on. But there are huge problems with this wide definition of 'function', because many of the things that happened to the DNA did not imply that the DNA had an actual job to do for the body, merely that it was subject to housekeeping chemical processes. Realising they had gone too far, some of the ENCODE team began to use smaller numbers when interviewed afterwards. One claimed only 20 per cent was functional, before insisting none the less that the term 'junk DNA' should be 'totally expunged from the lexicon' – which, as Dan Graur of the University of Houston and his colleagues remarked in a splenetic riposte in early 2013, thus invented a new arithmetic according to which 20 per cent is greater than 80 per cent.

If this all seems a bit abstruse, perhaps an analogy will help. The function of the heart, we would surely agree, is to pump blood. That is what natural selection has honed it to do. The heart does other things, such as add to the weight of the body, produce sounds and prevent the pericardium from deflating. Yet to call those the functions of the heart is silly. Likewise, just because junk DNA is sometimes transcribed or altered, that does not mean it has function as far as the body is concerned. In effect, the ENCODE team was arguing that grasshoppers are three times as complex, onions five times and lungfish forty times as complex, as human beings. As the evolutionary biologist Ryan Gregory put it, anyone who thinks he or she can assign a function to every letter in the human genome should be asked why an onion needs a genome that is about five times larger than a person's.

Who's resorting to a skyhook here? Not Ohno or Dawkins or Gregory. They are saying the extra DNA just comes about, there not being sufficient selective incentive for the organism to clear out its genomic attic. (Admittedly, the idea of junk in your attic that duplicates itself if you do nothing about it is moderately alarming!) Bacteria, with large populations and brisk competition to grow faster than their rivals, generally do keep their genomes clear of junk. Large organisms do not. Yet there is clearly a yearning that many people have to prefer an explanation that sees the spare DNA as having a purpose for us, not for itself. As Graur puts it, the junk critics have fallen prey to 'the genomic equivalent of the human propensity to see meaningful patterns in random data'.

Whenever I raised the topic of junk DNA in recent years I was astonished by the vehemence with which I was told by scientists and commentators that I was wrong, that its existence had been disproved. In vain did I point out that on top of the transposons, the genome was littered with 'pseudogenes' – rusting hulks of dead genes – not to mention that 96 per cent of the RNA transcribed from genes was discarded before proteins were made from the transcripts (the discards are 'introns'). Even though some parts of introns and pseudogenes are used in control sequences, it was clear the bulk was just taking up space, its sequence free to change without consequence for the body. Nick Lane argues that even introns are descended from digital parasites, from the period when an archeal cell ingested a bacterium and turned it into the first mitochondrion, only to see its own DNA invaded by selfish DNA sequences from the ingested bacterium: the way introns are spliced out betrays their ancestry as self-splicing introns from bacteria.

Junk DNA reminds us that the genome is built by and for DNA sequences, not by and for the body. The body is an emergent phenomenon consequent upon the competitive survival of DNA sequences, and a means by which the genome perpetuates itself. And though the natural selection that results in evolutionary change is very far from random, the mutations themselves are random. It is a process of blind trial and error.

Red Queen races

Even in the heart of genetics labs there is a long tradition of resistance to the idea that mutation is purely random and comes with no intentionality, even if selection is not random. Theories of directed mutation come and go, and many highly reputable scientists embrace them, though the evidence remains elusive. The molecular biologist Gabby Dover, in his book *Dear Mr Darwin*, tried to explain the implausible fact that some centipedes have 173 body segments without relying exclusively on natural selection. His argument was basically that it was unlikely that a randomly generated 346-legged centipede survived and bred at the expense of one with slightly fewer legs. He thinks some other explanation is needed for how the centipede got its segments. He finds such an explanation in 'molecular drive', an idea that remains frustratingly vague in Dover's book, but has a strong top–down tinge. In the years since Dover put forward the notion, molecular drive has sunk with little trace, following so many other theories of directed mutation into oblivion. And no wonder: if mutation is directed, then there would have to be a director, and we're back to the problem of how the director came into existence: who directed the director? Whence came this knowledge of the future that endowed a gene with the capacity to plan a sensible mutation?

In medicine, an understanding of evolution at the genomic level is both the problem and the solution. Bacterial resistance to antibiotics, and chemotherapeutic drug resistance within tumours, are both pure Darwinian evolutionary processes: the emergence of survival mechanisms through selection. The use of antibiotics selects for rare mutations in genes in bacteria that enable them to resist the drugs. The emergence of antibiotic resistance is an evolutionary process, and it can only be combated by an evolutionary process. It is no good expecting somebody to invent the perfect antibiotic, and find some way of using it that does not elicit resistance. We are in an arms race with germs, whether we like it or not. The mantra should always be the Red Queen's (from Lewis Carroll's *Through the Looking-Glass*): 'Now, here, you see, it takes all the running you can do, to keep in the same place. If you want to get somewhere else, you must run at least twice as fast as that!' The search for the next antibiotic must begin long before the last one is ineffective.

That, after all, is how the immune system works. It does not just produce the best antibodies it can find; it sets out to experiment and evolve in real time. Human beings cannot expect to rely upon evolving resistance to parasites quickly enough by the selective death of susceptible people, because our generation times are too long. We have to allow evolution within our bodies within days or hours. And this the immune system is designed to achieve. It contains a system for recombining different forms of proteins to increase their diversity and rapidly multiplying whichever antibody suddenly finds itself in action. Moreover, the genome includes a set of genes whose sole aim seems to be to maintain a huge diversity of forms: the major histocompatibility complex. The job of these 240 or so MHC genes is to present antigens from invading pathogens to the immune system so as to elicit an immune response. They are the most variable genes known, with one – HLA-B – coming in about 1,600 different versions in the human population. There is some evidence that many animals go to some lengths to maintain or enhance the variability further, by, for example, seeking out mates with different MHC genes (detected by smell).

If the battle against microbes is a never-ending, evolutionary arms race, then so is the battle against cancer. A cell that turns cancerous and starts to grow into a tumour, then spreads to other parts of the body, has to evolve by genetic selection as it does so. It has to acquire mutations that encourage it to grow and divide; mutations that ignore the instructions to stop growing or commit suicide; mutations that cause blood vessels to grow into the tumour to supply it with nutrients; and mutations that enable cells to break free and migrate. Few of these mutations will be present in the first cancerous cell, but tumours usually acquire another mutation – one that massively rearranges its genome, thus experimenting on a grand scale, as if unconsciously seeking to find a way by trial and error to acquire these needed mutations.

The whole process looks horribly purposeful, and malign.

The tumour is 'trying' to grow, 'trying' to get a blood supply, 'trying' to spread. Yet, of course, the actual explanation is emergent: there is competition for resources and space among the many cells in a tumour, and the one cell that acquires the most helpful mutations will win. It is precisely analogous to evolution in a population of creatures. These days, the cancer cells often need another mutation to thrive: one that will outwit the chemotherapy or radiotherapy to which the cancer is subjected. Somewhere in the body, one of the cancer cells happens to acquire a mutation that defeats the drug. As the rest of the cancer dies away, the descendants of this rogue cell gradually begin to multiply, and the cancer returns. Heartbreakingly, this is what happens all too often in the treatment of cancer: initial success followed by eventual failure. It's an evolutionary arms race.

The more we understand genomics, the more it confirms evolution.

5

The Evolution of Culture

And therefore to assume there was one person gave a name
To everything, and that all learned their first words from the same,
Is stuff and nonsense. Why should one human being from among
The rest be able to designate and name things with his tongue
And others not possess the power to do likewise? . . .

Lucretius, *De Rerum Natura*, Book 5, lines 1041–5

The development of an embryo into a body is perhaps the most beautiful of all demonstrations of spontaneous order. Our understanding of how it happens grows ever less instructional. As Richard Dawkins writes in his book *The Greatest Show on Earth*, 'The key point is that there is no choreographer and no leader. Order, organisation, structure – these all emerge as by-products of rules which are obeyed locally and many times over.' There is no overall plan, just cells reacting to local effects. It is as if an entire city emerged from chaos just because people responded to local incentives in the way they set up their homes and businesses. (Oh, hang on – that is how cities emerged too.)

Look at a bird's nest: beautifully engineered to provide protection and camouflage to a family of chicks, made to a consistent (but unique) design for each species, yet constructed by the simplest of instincts with no overall plan in mind, just a string of innate urges. I had a fine demonstration of this one year when a mistle thrush tried to build a nest on the metal fire escape outside my office. The result was a disaster, because each step of the fire escape looked identical, so the poor bird kept getting confused about which step it was building its nest on. Five different steps had partly built nests on them, the middle two being closest to completion, but neither fully built. The bird then laid two eggs in one half-nest and one in another. Clearly it was confused by the local cues provided by the fire-escape steps. Its nest-building program depended on simple rules, like 'Put more material in corner of metal step.' The tidy nest of a thrush emerges from the most basic of instincts.

Or look at a tree. Its trunk manages to grow in width and strength just as fast as is necessary to bear the weight of its branches, which are themselves a brilliant compromise between strength and flexibility; its leaves are a magnificent solution to the problem of capturing sunlight while absorbing carbon dioxide and losing as little water as possible: they are wafer-thin, feather-light, shaped for maximum exposure to the light, with their pores on the shady underside. The whole structure can stand for hundreds or even thousands of years without collapsing, yet can also grow continuously throughout that time – a dream that lies far beyond the capabilities of human engineers. All this is achieved without a plan, let alone a planner. The tree does not even have a brain. Its design and implementation emerge from the decisions of its trillions of single cells. Compared with animals, plants dare not rely on brain-directed behaviour, because they cannot run away from grazers, and if a grazer ate the brain, it would mean death. So plants can withstand almost any loss, and regenerate easily. They are utterly decentralised. It is as if an entire country's economy emerged from just the local incentives and responses of its people. (Oh, hang on . . .)

Or take a termite mound in the Australian outback. Tall, buttressed, ventilated and oriented with respect to the sun, it is a perfect system for housing a colony of tiny insects in comfort and gentle warmth – as carefully engineered as any cathedral. Yet there is no engineer. The units in this case are whole termites, rather than cells, but the system is no more centralised than in a tree or an embryo. Each grain of sand or mud that is used to construct the mound is carried to its place by a termite acting under no instruction, and with no plan in (no) mind. The insect is reacting to local signals. It is as if a human language, with all its syntax and grammar, were to emerge spontaneously from the actions of its individual speakers, with nobody laying down the rules. (Oh, hang on . . .)

That is indeed exactly how languages emerged, in just the same fashion that the language of DNA developed – by evolution. Evolution is not confined to systems that run on DNA. One of the great intellectual breakthroughs of recent decades, led by two evolutionary theorists named Rob Boyd and Pete Richerson, is the realisation that Darwin's mechanism of selective survival resulting in cumulative complexity applies to human culture in all its aspects too. Our habits and our institutions, from language to cities, are constantly changing, and the mechanism of change turns out to be surprisingly Darwinian: it is gradual, undirected, mutational, inexorable, combinatorial, selective and in some vague sense progressive.

Scientists used to object that evolution could not occur in culture because culture did not come in discrete particles, nor did it replicate faithfully or mutate randomly, like DNA. This turns out not to be true. Darwinian change is inevitable in any system of information transmission so long as there is some lumpiness in the things transmitted, some fidelity of transmission and a degree of randomness, or trial and error, in innovation. To say that culture 'evolves' is not metaphorical.

The evolution of language

There is an almost perfect parallel between the evolution of DNA sequences and the evolution of written and spoken language. Both consist of linear digital codes. Both evolve by selective survival of sequences generated by at least partly random variation. Both are combinatorial systems capable of generating effectively infinite diversity from a small number of discrete elements. Languages mutate, diversify, evolve by descent with modification and merge in a ballet of unplanned beauty. Yet the end result is structure, and rules of grammar and syntax as rigid and formal as you could want. 'The formation of different languages, and of distinct species, and the proofs that both have been developed through a gradual process, are curiously parallel,' wrote Charles Darwin in *The Descent of Man*.

This makes it possible to think of language as a designed and rule-based thing. And for generations, this was the way foreign languages were taught. At school I learned Latin and Greek as if they were cricket or chess: you can do this, but not that, to verbs, nouns and plurals. A bishop can move diagonally, a batsman can run a leg bye, and a verb can take the accusative. Eight years of this rule-based stuff, taught by some of the finest teachers in the land for longer hours each week than any other topic, and I was far from fluent – indeed, I quickly forgot what little I had learned once I was allowed to abandon Latin and Greek. Top–down language teaching just does not work well – it's like learning to ride a bicycle in theory, without ever getting on one. Yet a child of two learns English, which has just as many rules and regulations as Latin, indeed rather more, without ever being taught. An adolescent picks up a foreign language, conventions and all, by immersion. Having a training in grammar does not (I reckon) help prepare you for learning a new language much, if at all. It's been staring us in the face for years: the only way to learn a language is bottom–up.

Language stands as the ultimate example of a spontaneously

organised phenomenon. Not only does it evolve by itself, words changing their meaning even as we watch, despite the railings of the mavens, but it is learned, not taught. The prescriptive habit has us all tut-tutting at the decline of language standards, the loss of punctuation and the debasement of vocabulary, but it's all nonsense. Language is just as rule-based in its newest slang forms, and just as sophisticated as it ever was in ancient Rome. But the rules, now as then, are written from below, not from above.

There are regularities about language evolution that make perfect sense but have never been agreed by committees or recommended by experts. For instance, frequently used words tend to be short, and words get shorter if they are more frequently used: we abbreviate terms if we have to speak them often. This is good – it means less waste of breath, time and paper. And it is an entirely natural, spontaneous phenomenon that we remain largely unaware of. Similarly, common words change only very slowly, whereas rare words can change their meaning and their spelling quite fast. Again, this makes sense – re-engineering the word 'the' so it means something different would be a terrific problem for the world's English-speakers, whereas changing the word 'prevaricate' (it used to mean 'lie', it now seems mostly to mean 'procrastinate') is no big deal, and has happened quite quickly. Nobody thought up this rule; it is the product of evolution.

Languages show other features of evolutionary systems. For instance, as Mark Pagel has pointed out, biological species of animals and plants are more diverse in the tropics, less so near the poles. Indeed, many circumpolar species tend to have huge ranges, covering the whole of an ecosystem in the Arctic or Antarctic, whereas tropical rainforest species might be found in just one small area – a valley or a mountain range or on an island. The rainforest of New Guinea is a menagerie of millions of different species with small ranges, while the tundra of Alaska is home to a handful of species with vast ranges. This is true of plants, insects, birds, mammals, fungi. It's a sort of iron rule of ecology: that there will be more species, but with smaller ranges, near the equator, and fewer species, but with larger ranges, near the poles.

And here is the fascinating parallel. It is also true of languages. The native tongues spoken in Alaska can be counted on one hand. In New Guinea there are literally thousands of languages, some of which are spoken in just a few valleys and are as different from the languages of the next valley as English is from French. Even this language density is exceeded on the volcanic island of Gaua, part of Vanuatu, which has five different native languages in a population of just over 2,000, despite being a mere thirteen miles in diameter. In forested, mountainous tropical regions, human language diversity is extreme.

One of Pagel's graphs shows that the decreasing diversity of languages with latitude is almost identical to the decreasing diversity of species with latitude. At present neither trend is easily explained. The great diversity of species in tropical forests has something to do with the greater energy flowing through a tropical ecosystem with plenty of warmth and light and water. It may also have something to do with the abundance of parasites. Tropical creatures are subjected to a constant barrage of parasitic invasions, and being an abundant creature makes you more of a target, so there is an advantage to rarity. And it may reflect a lower extinction rate in a more climatically equable zone. As for languages, the need to migrate with the seasons must homogenise the linguistic diversity of extremely seasonal landscapes, in contrast to tropical ones, where populations can fragment into smaller groups and each can survive without moving. But whatever the explanation, the phenomenon illustrates the way human languages evolve automatically. They are clearly human products, but they are not consciously designed.

Moreover, by studying the history of languages, Pagel finds that when a new language diverges from an ancestral language, it appears to change very rapidly at first. The same seems to be true of species. When a geographical subset of a species becomes isolated it evolves very rapidly at first, so that evolution by natural selection seems to happen in bursts, a phenomenon known as punctuated equilibrium. There are intensely close parallels between the evolution of languages and of species.

The human revolution was actually an evolution

Some time around 200,000 years ago, in a part of Africa but not elsewhere, human beings began to change their culture. We know this because the archaeological record is clear that a great transformation came over the species, known as the 'human revolution'. After more than a million years of making simple stone tools to just a few designs, these Africans began making lots of different types of tool. At first the change was local, gradual and ephemeral, so the word revolution is misleading. But then the tool changes began to appear more frequently, more strongly and more persistently. By 65,000 years ago the people with the new tool sets had begun to spill out of Africa, most probably across the narrow strait at the southern end of the Red Sea, and had begun a comparatively rapid colonisation of the Eurasian continent, displacing – and very occasionally mating with – the native hominids that were already there, such as the Neanderthals in Europe and the Denisovans in Asia. These new people had something special: they were not prisoners of their ecological niche, but could change their habits quite easily if prey disappeared, or better opportunities arose. They reached Australia and quickly filled that challenging continent. They reached Europe, then in the grip of an ice age, and displaced the superbly adapted big-game-hunting Neanderthals. They eventually spilled into the Americas, and within an evolutionary eye-blink peopled every ecosystem from Alaska to Cape Horn, from the rainforest to the desert.

What sparked the human revolution in Africa? It is an almost impossibly difficult question to answer, because of the very gradual beginning of the process: the initial trigger may have been very small. The first stirrings of different tools in parts of east Africa seem to be up to 300,000 years old, so by modern standards the change was happening with glacial slowness. And that's a clue. The defining feature is not culture, for plenty of animals have culture, in the sense of traditions that are passed on by learning. The defining feature is cumulative culture – the capacity to add innovations without losing old habits. In this sense, the human revolution was not a revolution at all, but a very, very slow cumulative change, which steadily gathered pace, accelerating towards today's near-singularity of incessant and multifarious innovation.

It was cultural evolution. I think the change was kicked off by the habit of exchange and specialisation, which feeds upon itself – the more you exchange, the more value there is in specialisation, and vice versa – and tends to breed innovation. Most people prefer to think it was language that was the cause of the change. Again, language would build upon itself: the more you can speak, the more there is to say. The problem with this theory, however, is that genetics suggests Neanderthals had already undergone the linguistic revolution hundreds of thousands of years earlier – with certain versions of genes related to languages sweeping through the species. So if language was the trigger, why did the revolution not happen earlier, and to Neanderthals too? Others think that some aspect of human cognition must have been different in these first 'behaviourally modern humans': forward planning, or conscious imitation, say. But what caused language, or exchange, or forethought, to start when and where it did?

Almost everybody answers this question in biological terms: a mutation in some gene, altering some aspect of brain structure, gave our ancestors a new skill, which enabled them to build a culture that became cumulative. Richard Klein, for instance, talks of a single genetic change that 'fostered the uniquely modern ability to adapt to a remarkable range of natural and social circumstance'. Others have spoken of alterations in the size, wiring and physiology of the human brain to make possible everything from language and tool use to science and art. Others suggest that a small number of mutations, altering the structure or expression of developmental regulatory genes, were what triggered a cultural explosion. The evolutionary geneticist Svante Pääbo says: 'If there is a genetic underpinning to this cultural and technological explosion, as I'm sure there is . . .'

I am not sure there is a genetic underpinning. Or rather, I think they all have it backwards, and are putting the cart before the horse. I think it is wrong to assume that complex cognition is what makes human beings uniquely capable of cumulative cultural evolution. Rather, it is the other way around. Cultural evolution drove the changes in cognition that are embedded in our genes. The changes in genes are the consequences of cultural changes. Remember the example of the ability to digest milk in adults, which is unknown in other mammals, but common among people of European and east African origin. The genetic change was a response to the cultural change. This happened about 5,000–8,000 years ago. The geneticist Simon Fisher and I argued that the same must have been true for other features of human culture that appeared long before that. The genetic mutations associated with facilitating our skill with language – which show evidence of 'selective sweeps' in the past few hundred thousand years, implying that they spread rapidly through the species – were unlikely to be the triggers that caused us to speak; but were more likely the genetic responses to the fact that we were speaking. Only in a language-using animal would the ability to use language more fluently be an advantage. So we will search in vain for the biological trigger of the human revolution in Africa 200,000 years ago, for all we will find is biological responses to culture. The fortuitous adopting of a habit, through force of circumstance, by a certain tribe might have been enough to select for genes that made the members of that tribe better at speaking, exchanging, planning or innovating. In people, genes are probably the slaves, not the masters, of culture.

Music, too, evolves. To a surprising extent, it changes under its own steam, with musicians carried along for the ride. Baroque begets classical begets romantic begets ragtime begets jazz begets blues begets rock begets pop. One style could not emerge without the previous style existing. There are hybridisation events along the way: African traditional music mates with blues to produce jazz. Instruments change, but mainly as a result of descent with modification from other instruments, not by *de novo* invention. The piano is the descendant of the harpsichord, which shares an ancestor with the harp. The trombone is the daughter of the trumpet and the cousin of the horn. The violin and the cello are modified lutes. Just as Mozart could not have written what he did if Bach and his like had not written what they did, nor could Beethoven have written his music without drawing upon Mozart. Technology matters, but so do ideas: Pythagoras's discovery of the octave scale was a crucial moment in the history of music. So was syncopation. The invention of the amplified electric guitar made small groups able to entertain large ones as easily as orchestras once could. The point is that there was an inexorable inevitability about the gradual progress of music. It could not stop changing as each generation of musicians learned and experimented with music.

The evolution of marriage

One of the characteristics of evolution is that it produces patterns of change that make sense in retrospect, but that came about without even a hint of conscious design. Take the human mating system. The emergence, fall, rise and fall again of marriage over the last few thousand years constitute a fine example of this pattern. I am not talking about the evolution of mating instincts, but the history of cultural marriage habits.

The instincts are there, sure enough. Human mating patterns plainly still reflect ingrained genetic tendencies honed in the African savannah over millions of years. Judging by the modest difference in size and strength between men and women, we are clearly not designed for pure polygamy of the gorilla kind, where gigantic males compete for ownership of stable harems of females, killing the predecessor male's babies when they succeed. On the other hand, judging by the modest size of our testicles, we are not designed for the sexual free-for-all of chimpanzees and bonobos, where promiscuous females (in what is probably an instinctive bid to prevent infanticide) ensure that most male-to-male competition happens between sperm rather than between individuals, blurring paternity. We are like neither of these. Hunter-gatherer societies, it turned out once we got to study them starting in the 1920s, are mainly monogamous. Males and females form exclusive pair bonds, and if either sex desires sexual variety it largely seeks it in secret. Monogamous pair bonding, with fathers closely involved in providing for offspring, seems to be the peculiar human pattern that men and women adopted for most of the past few million years. This is unusual among mammals, much more common among birds.

But as soon as farming came along, 10,000 years ago, powerful men were able to accumulate the resources to buy off and intimidate other men, and to attract low-status women into harems. From ancient Egypt to the Inca empire, from the farming cultures of west Africa to the herding societies of central Asia, polygamy became the norm, whatever the instincts. This suited powerful men and low-status women (who could be pampered as the ninth wife of a rich man, rather than starving as the only wife of a poor man), but it was not such a good deal for low-status men, who remained single, or high-status women, who had to share their partners' attention. If only to try to satisfy the low-status men, societies that allowed widespread polygamy tended to be very violent towards their neighbours. This was especially true of pastoral societies reliant on sheep, goats or cattle, whose wealth was mobile and showed scale economies: a thousand sheep are not much harder to look after than five hundred. So herders from Asia and Arabia not only experienced chronic violence, but kept erupting into Europe, India, China and Africa to kill men and abduct women. People like Attila, Genghis, Kublai, Tamerlane, Akbar and many others. Their habit was to conquer a country, kill its men, children and old women, and take its young women as concubines. Genghis himself fathered thousands of children, but his followers were not far behind.

The point is that the emergence of human polygamy among herders makes economic and ecological sense in retrospect, but that does not mean it was designed for the purpose by some clever inventor. The rationale never existed inside the heads of those inventing it – it's what Daniel Dennett calls a 'free-floating rationale'. It was an adaptive, evolutionary consequence of a certain set of selective conditions.

Polygamy took a different form in agrarian societies, like Egypt, west Africa, Mexico and China. High-status men had more wives than low-status men but, emperors apart, not to the same extremes as in pastoral societies. Often, as in west Africa, rich men were like parasites living off the hard work of a group of women they called wives. In exchange for protection against other men, the women got to live on, and cultivate, the polygamous husband's land.

However, in some of these settled civilisations trading cities grew up, and these generated a wholly new selective pressure – towards monogamy, fidelity and marriage. You can glimpse this transition in the difference between the *Iliad* – full of infighting between polygamist men – and the *Odyssey* – which features the story of a virtuous Penelope waiting for her (mostly) faithful Odysseus. The tradition of the high-born, virtuous woman holding out for proper marriage rather than submitting to the indignity of concubinage appears also in the Roman founding myth of the rape of Lucretia, where it is intimately connected with the birth of the republic itself and the overthrow of kings – the implication being that what brought down the kings was their overbearing tendency to steal other men's women, ensuring resentment among other men and among women.

This transition to monogamy is a big theme of Christianity and an incessant preoccupation of the early Church fathers, though not all early saints recommended monogamy. In the teachings of Jesus they found cause to insist on each man taking one wife and staying with her through good times and bad. Marriage, Christ was said to have taught, was a holy state – two became 'one flesh'. The winners from the re-emergence of monogamy in late antiquity would have been the high-born women, who got to monopolise their husbands, and the much more numerous low-born men, who got to have sex at all. It was in appealing to these low-born men that early Christians struck evangelical gold.

Not that polygamy disappeared altogether. The battle between polygamous aristocrats (attractive to low-born women as an escape route from starvation) and the bourgeois virtues preferred by their prime wives and their yeomen subjects was played out throughout the Dark Ages, the Middle Ages and early modern history. Sometimes one side was in the ascendant, sometimes the other. Under Oliver Cromwell's puritans in seventeenth-century England, monogamy prevailed. Under Charles II, polygamy returned – unofficially. A brief biography of a famous soldier, Prince Maurice de Saxe, begins thus: 'Eldest of the 354 acknowledged illegitimate children of Frederick Augustus, Elector of Saxony, King of Poland, Maurice of Saxony, the prodigious Marshal, was born October 28th 1696 . . .' Maurice himself was no slouch in the sexual stakes, fathering his first child as a fifteen-year-old during the siege of Tournai, and then squandering his wife's fortune on maintaining 'his regiment of horse and his legion of mistresses'.

The resentment that such behaviour generated is not hard to imagine, and in the relative freedom from feudal duties of the market towns, the sons and daughters of the bourgeoisie did not put up with it. In the eighteenth century it is no accident that one of the dominant themes of popular literature – as exemplified by the stories of Figaro in France, Pamela (in the novel by Richardson) in England – is the rebellion of a man of modest means against the *droit du seigneur* of a nobleman. Monogamy eventually conquered even the nobility with the rise of the merchant class, and by the nineteenth century Queen Victoria had tamed the appetites even of royal men to the point where every man had at least to pretend that he was the faithful, attentive and lifelong devotee of one woman. It is no accident, says William Tucker in his brilliant book *Marriage and Civilization*, that on the whole peace comes to Europe as a result. Peace, that is, except where societies continue to be based on polygamy, such as much of the Muslim world, or where polygamy was suddenly reinvented, as in the Church of Jesus Christ of Latter Day Saints. The Mormons' polygamy caused huge resentment among neighbours, as well as tensions among Saints, and cycles of terrible violence followed them on their peregrinations all the way to Utah. It culminated in the Mountain Meadow massacre of 1857, carried out in revenge for the killing by an enraged husband of a Mormon who had lured the man's wife into joining his harem. The violence died down only with the outlawing of polygamy in 1890. (Unofficial polygamy persists to this day in a very few Mormon fundamentalist communities.)

The foremost anthropologists of cultural evolution, Joe Henrich, Rob Boyd and Pete Richerson, have argued in an influential paper called 'The Puzzle of Monogamous Marriage' that the spread of monogamy in the modern world can best be explained by its beneficial effects on society. That is to say, not that clever men sat around a table and decided upon a policy of monogamy in order to bring peace and cohesion, but more likely that it was a case of cultural evolution by Darwinian means. Societies that chose 'normative monogamy', or an insistence upon sex within exclusive marriage, tended to tame their young men, improve social cohesion, balance the sex ratio, reduce the crime rate, and encourage men to work rather than fight. This made such societies more productive and less destructive, so they tended to expand at the expense of other societies. That, the three anthropologists think, explains the triumph of monogamy, which reaches its apogee in the perfect nuclear family of 1950s America, with Dad going out to work and Mom at home cleaning, cooking and looking after the kids.

Incidentally, Tucker points out a fascinating episode in the history of wage bargaining. In the early twentieth century there was a remarkably successful campaign to force employers to pay men higher wages precisely so that their wives would not have to work: the 'family wage' movement. Social reformers, far from wanting women to join the workforce, favoured the opposite: allowing them to leave the workforce and spend time with their children, supported by a better-paid husband. The argument they advanced was that if employers would only pay more, then working-class women would be able to join middle-class women in not having to work outside the home.

Then with the rise of the welfare state, in the late twentieth century monogamy began to break down again. Once the man's role of breadwinner is replaced by a welfare payment, it is an empirical fact that many women increasingly begin to think that monogamy is a form of indentured servitude they could do without. Some parts of society have abandoned marriage and adopted the practice of single motherhood, serviced by wandering, polygamous men. Perhaps this was because increasingly women began to see the solidarity of their feminist sisters as a more permanent and progressive option for the social support of young mothers. Either that or men decided they need no longer hang around to see their children safely into adulthood. Or a bit of both. Whatever your preferred explanation for the breakdown of marriage in recent years, there is no doubt that it is an institution that is evolving before our very eyes, and that it will look very different by the end of this century. Marriage is not redesigned; it evolves. We don't really notice it happening till afterwards. But the change is far from random.

The evolution of cities

Once you start noticing evolution at work in human affairs, you see it everywhere. Take cities. Between 1740 and 1850 Britain became the most urbanised country in the world, in a wholly unplanned way. Manchester, Birmingham, Leeds and Bristol swelled from little towns to great cities. The elegance of Bath and Cheltenham, the West End of London and Bloomsbury, the New Town in Edinburgh, and Grainger Town in Newcastle-upon-Tyne – all were built in this period. This was not a creation of the state or public authorities. All of it happened in a society with no apparatus of planning laws and regulatory bodies, no public building regulations, no zoning or land-use laws, no direct public action to supply housing or urban services.

Only in the second half of the nineteenth century was there a move to greater state control. The urban growth of the early period was driven by private initiative and speculation, directed by property rights and private contracts, and shaped and determined by decentralised market forces. This urbanisation was orderly but unplanned. It was evolutionary.

Cities first emerged in the Bronze Age, when pack animals and boats enabled people to bring food in sufficient quantities to settlements bigger than villages. They grew larger in the Iron Age, when wheeled carts and sailing ships led to bigger markets. They sprawled outward when horse-drawn omnibuses and then steam trains gave people the option of longer commutes. They expanded rapidly when cars and trucks drew people in growing numbers to the biggest urban areas. And then they began to change from centres of production to centres of consumption. In America as a whole, nearly twice as many people work in grocery stores as in restaurants. In Manhattan, nearly five times as many work in restaurants as in grocery stores. When corrected for age, education and marital status, city-dwellers are 44 per cent more likely to visit a museum and 98 per cent more likely to go to a movie theatre than rural Americans.

The sociologist Jane Jacobs was the first to realise that the density of urban living, 'far from being an evil, is the source of its vitality' (in John Kay's words). In her successful opposition to New York's city planners with their utopian schemes, she championed the unplanned, organic nature of the cities that people love, in contrast to the sterile spaces of planned cities like Brasilia, Islamabad or Canberra. As Nassim Taleb quips, nobody would buy a pied-à-terre in Brasilia the way they would in London.

Today, the most successful cities, like London, New York and Tokyo, are places of fancy food, entertainment, mating arenas (sorry – clubs) and opportunity for the aspiring poor. From Rio to Mumbai cities are the engines of prosperity, the places where people make the transition from poverty to comfort and even wealth. And the 'death of distance' engendered by the internet and the mobile phone, far from encouraging people to retreat to isolated idylls in Montana or the Gobi desert, is having exactly the opposite effect. Now that we can work anywhere, the anywhere we mostly want – at least when we are young – is the densest, most high-rise, most hectic of spots. And we are prepared to pay a premium for it. Cities that encourage tall residential buildings in their centres, like Hong Kong or Vancouver, thrive, while those that insist on low-rise profiles, like Mumbai, struggle. The point is that these are not trends that human beings have chosen consciously as policy. The continuing evolution of the city is an unconscious and inexorable momentum.

The same process is continuing all over the world. There is, as Edward Glaeser has observed, an almost perfect correlation between prosperity and urbanisation: the more urbanised a country, the richer it is. If you divide the world into those countries where the majority live in cities, and those where the majority live in the countryside, you find that the former are four times as wealthy, in terms of average income, as the latter. As more and more people move into cities and they grow larger and larger, some scientists have begun to notice that cities themselves evolve in predictable ways. There is a spontaneous order in the way they grow and change. The most striking of these regularities is the 'scaling' that cities show – how their features change with size. For example, the number of petrol stations increases at a consistently slower rate than the population of the city. There are economies of scale, and this pattern is the same in every part of the world. The same is true of electrical networks. So it does not matter what the policy of the country, or the mayor, is. Cities will converge on the same patterns of growth wherever they are. In this they are very like bodies. A mouse burns more energy, per unit of body weight, than an elephant; a small city burns proportionately more motor fuel than a large one. Like cities, bodies get more efficient in their energy consumption the larger they grow. There is also a consistent 15 per cent saving on infrastructure cost per head for every doubling of a city's population size.

The opposite is true of economic growth and innovation – the bigger the city, the faster these increase. Doubling the size of a city boosts income, wealth, number of patents, number of universities, number of creative people, all by approximately 15 per cent, regardless of where the city is. The scaling is, in the jargon, 'superlinear'. Geoffrey West of the Santa Fe Institute, who discovered this phenomenon, calls cities 'supercreative'. They generate a disproportionate share of human innovation; and the bigger they are, the more they generate. The reason for this is clear, at least in outline. Human beings innovate by combining and recombining ideas, and the larger and denser the network, the more innovation occurs. Once again, notice that this is not policy. Indeed, nobody was aware of the supercreative effect of cities until very recently, so no policy-maker could aim for it. It's an evolutionary phenomenon.

It is one of the reasons that cities hardly ever die. Apart from Detroit today and Sybaris in ancient times, there are few examples of cities that even shrink, let alone vanish in the way that – say – companies do all the time.

The evolution of institutions

Some species evolve into new forms very rapidly, while others stay the same for hundreds of millions of years. These are known as living fossils. The coelacanth is a good example – a deep-sea fish very similar in form to its ancestors of 400 million years ago. The same can be true of cultural evolution: some institutions change very rapidly, while others retain their form for centuries. Britain is as modern a country as any. It has nearly all modern technologies in abundance, it contributes more than most to scientific discovery, it has moved with the times in social terms, from legalising gay marriage to appointing women bishops. But Britain's political institutions have changed surprisingly little in three centuries. As the sociologist Garry Runciman points out in his book *Very Different, But Much the Same*, if Daniel Defoe, who observed and wrote about British life in the early eighteenth century, were to return to London today he would find some things astonishingly unchanged. Once he had got used to aeroplanes, toilets, cars, phones, photography, pensions, the internet, religious diversity, vaccines, women lawyers, electricity, and vastly higher living standards especially for the poor, he would be able to understand British politics very easily. There is still a hereditary monarch who is head of the established Church of England, and an elected House of Commons and an appointed House of Lords. There are parties, factions, scandals and systems of patronage that are distinctly Hanoverian in outline at least. In those days Britain had the population and the income per head roughly those of modern Togo.

Runciman is keen on the theory of cultural evolution, the notion that new ways of doing things emerge gradually and survive if they suit society, rather than being imposed by grand design. But why do technologies, clothing, language, music and economic activity change so fast, while political institutions change so slowly? In the stream of cultural evolution, British institutions are cultural coelacanths – living fossils that have stayed much the same while the world changed around them. Certainly, Britain stands out in this respect. Most other countries have changed their political institutions far more in the last three centuries after revolutions, wars or the gaining of independence. But everywhere, political institutions show a tendency to change much more slowly than the society around them, and when they do change, they do so with painful and traumatic lurches, called revolutions. China today has the economy of a twenty-first-century economic superpower with a political regime little changed since the 1950s.

Is this slow evolution in political institutions down to the concentration or the dispersion of power? Too many vested interests in the status quo or the fear of change among an elite? I am not quite sure of the answer. It is certainly hard to get people to vote for constitutional change. When allowed to choose products or services in a market, they are positively mad for new ideas. But when asked in referendums to agree on new political forms, they (in Hilaire Belloc's words) 'Always keep a hold of nurse, for fear of finding something worse.'

Cities, marriage, language, music, art – these manifestations of culture all change in regular and retrospectively predictable ways, but in ways that nobody did predict, let alone direct. They evolve.

6

The Evolution of the Economy

For you will find that everything for which we have a name
Is either a quality of the two, or consequence of the same.
A quality is what, without obliterating shock,
Can never be separated and removed: as weight to rock,
As heat to flame, wet to water, the ability to touch
To every substance, intangibility to the void. But such
As slavery, penury and riches, freedom, war and peace,
Whatever comes and goes while natures stay unchanging, these
We rightly tend to term as 'consequences' or 'events'.

Lucretius, *De Rerum Natura*, Book 1, lines 449–57

Depending on whose estimate you choose, and how you correct
for inflation, the average person alive in the world today earns
in a year between ten and twenty times as much money, in real
terms, as the average person earned in 1800. Or rather, he or
she can afford ten or twenty times as many goods or services.
Call it, as the economic historian Deirdre McCloskey does, the
'great enrichment'. She says it is the 'main fact or finding of economic history'. Indeed, says McCloskey, depending on how you
allow for improvements in things like steel girders, plate glass
and medicine, the standard of living could have risen as much

as a hundred times just since 1950 in a place like Hong Kong. At the rate the world economy is growing – and it has shown no signs of deceleration – the average human being may be earning up to sixteen times as much again in 2100 as he or she does today, according to the OECD: that's $175,000 a year in today's money. The Great Recession of 2008–09 was just a brief blip in global terms: one year when the global economy shrank by less than 1 per cent before growing by 5 per cent the next.

By far the lion's share of this improvement went (and still goes) to ordinary workers and the poor. As McCloskey puts it, although the rich got richer, 'millions more have gas heating, cars, smallpox vaccinations, indoor plumbing, cheap travel, rights for women, lower child mortality, adequate nutrition, taller bodies, doubled life expectancy, schooling for their kids, newspapers, a vote, a shot at university and respect'. Global inequality is currently falling fast as people in poor countries get richer quicker than people in rich countries. The proportion of the world population living on $1.25 a day, corrected for inflation, has gone from 65 per cent in 1960 to 21 per cent today.

Surprising as it may seem, the cause of the great enrichment is still unknown. That is to say, there are plenty of theories about why incomes started growing so rapidly in some parts of the world in the early nineteenth century, and this then spread to the rest of the world, and – despite repeated predictions that it would stop – they just keep on growing today. But none of these theories commands universal allegiance. Some credit institutions, others ideas, others individuals, others the harnessing of energy, yet others luck. They all agree on two things, however: nobody planned this, and nobody expected it. Prosperity emerged despite, not because of, human policy. It developed inexorably out of the interaction of people by a form of selective progress very similar to evolution. Above all, it was a decentralised phenomenon, achieved by millions of individual decisions, mostly in spite of the actions of rulers. Indeed, it is possible to argue, as Daron Acemoglu and James Robinson do, that countries like Britain and the United States grew rich precisely because their citizens overthrew the elites who monopolised power. It was the wider distribution of political rights that made government account-able and responsive to citizens, allowing the great mass of people to take advantage of economic opportunities.

Human action, but not human design

The great enrichment was an evolutionary phenomenon. Let's return to the late eighteenth century, when Britain stands on the brink of this great enrichment, and revisit that great thinker about the general theory of evolution, Adam Smith. In 1776 Smith published his second book, *The Wealth of Nations*. In it he set out to champion a different evolutionary idea from that which he had set out in his *Theory of Moral Sentiments*. If God was not the cause of morality, was government the cause of pros-perity? In Smith's day, commerce was a tightly regulated business, with joint-stock companies chartered specifically and exclusively by the state to have monopolies, and mercantilist trade policies designed to promote certain kinds of foreign exports, not to mention professions strictly licensed by the government. In the cracks between the paving stones of regulation and *dirigisme* individuals could buy and sell, but pretty well nobody thought this was the source of prosperity. Wealth meant accumulating precious things.

The 'physiocrats' in France had at least begun to suggest that productive work was the source of wealth, not heaps of gold. From François Quesnay, leader of the physiocrats, whom he met in 1766, Smith picked up the idea that mercantilist direction of trade was a mistake, as was government grabbing all the revenue of trade to spend on ruinous wars and futile luxuries: their cry was '*Laissez faire et laissez passer, le monde va de lui même!*' (Let do and let pass, the world goes on by itself!). But the physio-crats insisted, strangely, that the only kind of productive work was farming. Manufacture and services were wasteful frittering.

Smith said instead that the 'annual produce of the land and labour of the society' was what counted. Today we call that GDP.

So becoming more prosperous means the same as becoming more productive – growing more wheat, making more tools, serving more customers. And the 'greatest improvement in the productive power of labour', Smith argued, 'seems to have been the effects of the division of labour'. If the farmer supplies food to the ironmonger in exchange for tools, then both are more productive, because the first does not have to stop work and make a tool badly, while the latter does not have to stop work to till a field badly. Specialisation, accompanied by exchange, is the source of economic prosperity.

Here, in my own words, is what a modern version of Smithism claims. First, the spontaneous and voluntary exchange of goods and services leads to a division of labour in which people specialise in what they are good at doing. Second, this in turn leads to gains from trade for each party to a transaction, because everybody is doing what he is most productive at and has the chance to learn, practise and even mechanise his chosen task. Individuals can thus use and improve their own tacit and local knowledge in a way that no expert or ruler could. Third, gains from trade encourage more specialisation, which encourages more trade, in a virtuous circle. The greater the specialisation among producers, the greater is the diversification of consumption: in moving away from self-sufficiency people get to produce fewer things, but to consume more. Fourth, specialisation inevitably incentivises innovation, which is also a collaborative process driven by the exchange and combination of ideas. Indeed, most innovation comes about through the recombination of existing ideas for how to make or organise things.

The more people trade and the more they divide labour, the more they are working for each other. The more they work for each other, the higher their living standards. The consequence of the division of labour is an immense web of cooperation among strangers: it turns potential enemies into honorary friends. A woollen coat, worn by a day labourer, was (said Smith) 'the produce of a great multitude of workmen. The shepherd, the sorter of the wool, the wool-comber or carder, the dyer, the scribbler, the spinner, the weaver, the fuller, the dresser . . .' In parting with money to buy a coat, the labourer was not reducing his wealth. Gains from trade are mutual; if they were not, people would not voluntarily engage in trade. The more open and free the market, the less opportunity there is for exploitation and predation, because the easier it is for consumers to boycott the predators and for competitors to whittle away their excess profits. In its ideal form, therefore, the free market is a device for creating networks of collaboration among people to raise each other's living standards, a device for coordinating production and a device for communicating information about needs through the price mechanism. Also a device for encouraging innovation. It is the very opposite of the rampant and selfish individualism that so many churchmen and others seem to think it is. The market is a system of mass cooperation. You compete with rival producers, sure, but you cooperate with your customers, your suppliers and your colleagues. Commerce both needs and breeds trust.

Imperfect markets are better than no markets

Few would disagree with this formulation, but equally few would accept that the ideal is ever realised in practice. And that, churchmen aside, is where all the disagreement about markets comes from. Fine in theory, useless in practice – so goes the verdict of most right-thinking people on the topic of markets.

The question then becomes whether commerce only works if it is perfect. Are semi-free markets better than none? The economist William Easterly is in no doubt that the invisible hand is not Utopia: 'It is the process of driving out of business the incompetent in favour of the mediocre, the mediocre in favour of the good, and the good in favour of the excellent.' A glance at economic history makes clear that countries run by and in the interests of merchants have not been perfect, but they have always been more prosperous, peaceful and cultured than countries run by despots. Phoenicia versus Egypt; Athens versus Sparta; the Song Chinese versus the Mongols; Italian city states versus Charles V's Spain; the Dutch republic versus Louis XIV's France; a nation of shopkeepers (England) versus Napoleon; modern California versus modern Iran; Hong Kong versus North Korea; Germany in the 1880s versus Germany in the 1930s.

There is no longer much doubt that free commerce has a better economic or humanitarian record than command-and-control government. The examples just keep rolling in. Take the history of Sweden, for instance. Contrary to conventional wisdom, Sweden did not become wealthy as a result of having a big government imposing social democracy. When it liberalised a feudal economy and strongly embraced Smithian free trade and free markets in the 1860s, the result was rapid growth and the spawning of great enterprises over the next fifty years, including Volvo and Ericsson (companies that have since evolved new products). When it expanded government hugely in the 1970s, the result was currency devaluation, stagnation and slow growth, culminating in a full-blown economic crisis in 1992 and a rapid fall in the country's relative standing in the world's economic league table. When it cut taxes, privatised education and liberalised private healthcare in the 2000s, it rediscovered growth.

To argue that free commerce leads to more prosperity than government planning is not, of course, to argue that all government should be abolished. There is a vital role for government to play in keeping the peace, enforcing the rules and helping those who need help. But that is not the same as saying government should plan and direct economic activity. Likewise, for all its virtues, commerce is not perfect. It has a habit of encouraging wasteful and damaging extravagances, not least because it leads to the marketing of signals for conspicuous consumption.

The central feature of commerce, and the thing that distinguishes it from socialist planning, is that it is decentralised. No central direction is required to tell the economy how many woollen coats, laptops or cups of coffee are needed. Indeed, when somebody does try to do so, the result is a miserable mess. Or North Korea. Prices, if allowed free to rise or fall, will gravitate under competition towards the cost of production, as demand matches supply. Suppliers will direct their efforts to the products most valued at any one time, driving down price and satisfying the most intense demand. The system is run by the decisions of millions of individuals.

In this way, prosperity, when it grows at all, grows entirely organically, without any direction from above. The division of labour has emerged, uninvited, within society. It has evolved. It is stimulated by our natural willingness to trade. And 'the propensity to truck, barter, and exchange one thing for another', in Smith's famous phrase, comes naturally to human beings, but not to other animals: 'Nobody ever saw a dog make fair and deliberate exchange of a bone with another dog.' So it is this propensity, if encouraged, that will cause prosperity to increase. Government's role is to let it happen, not to direct it.

The central problem for systems of command and control, whether fascist, communist or socialist, is the knowledge problem. As champions of free enterprise from Frédéric Bastiat to Friedrich Hayek have pointed out, the knowledge required to organise human society is bafflingly voluminous. It cannot be held in a single human head. Yet human society is organised, none the less. As Bastiat put it in his 1850 *Economic Harmonies*, how would one even contemplate setting out to feed Paris, a city with hordes of people with myriad tastes? It is impossible. Yet it happens, without fail, every day (and Paris has a still vaster population today, with more eclectic taste in food). There is a close parallel with evolution here. The feeding of Paris and the working of the human eye are equally complex manifestations of order. But in neither case is there a central commanding intelligence. The knowledge is dispersed among millions of people/ genes. It is decentralised. As so often, Smith got there first, saying in *The Wealth of Nations*: 'The sovereign is completely discharged from a duty, in the attempting to perform which he must always be exposed to innumerable delusions, and for the proper performance of which no human wisdom or knowledge could ever be sufficient; the duty of superintending the industry of private people, and of directing it towards the employments most suitable to the interest of the society.'

Invisible hands

This decentralised emergence of order and complexity is the essence of the evolutionary idea that Adam Smith crystallised in 1776. In his famous metaphor, Smith made the guiding hand invisible: each person 'intends only his own security; and by directing that industry in such a manner as its produce may be of the greatest value, he intends only his own gain, and he is in this, as in many other cases, led by an invisible hand to promote an end which was no part of his intention'. Yet when Smith wrote his *Wealth of Nations*, there was little good evidence for his central idea that free exchange of goods and services would produce general prosperity. Up until the late eighteenth century much wealth creation had been by plunder in one form or another, and there was nothing remotely resembling a free-market government in power anywhere in the world.

Yet in the decades that followed the book's publication, Britain in particular (and then much of Europe and North America) played out an extraordinary story of rising living standards, falling inequality and declining violence – thanks largely to the partial and hesitant following of Smith's recipe. Sceptics might argue that the accumulation of plundered capital from the empire was the source of that wealth, but this is plainly nonsense. As Smith so clearly saw, colonies were mostly a drain and a military distraction. Nor can capital explain the sheer scale of what happened to living standards. As Deirdre McCloskey puts it, in the great enrichment of the past two hundred years average income in Britain went from about $3 a day to about $100 a day in real terms. That simply cannot be achieved by capital accumulation, which is why she (and I) refuse to use the misleading, Marxist word 'capitalism' for the free market. They are fundamentally different things.

Adam Smith is no paragon. He got plenty wrong, including his clumsy labour theory of value, and he missed David Ricardo's insight about comparative advantage, which explains why even a country (or person) that is worse than its trading partner at making everything will still be asked to supply something, the thing it or he is least bad at making. But the core insight that he had, that most of what we see in society is (in Adam Ferguson's words) the result of human action but not of human design, remains true to this day – and under-appreciated. This is true of language, of morality and of the economy. The Smithian economy is a process of exchange and specialisation among ordinary people. It is an emergent phenomenon.

Diminishing returns?

The really big thing that both Smith and Ricardo – and Robert Malthus and John Stuart Mill and all the other British political economists of the time – missed, however, was that they were living through the Industrial Revolution. They had no conception that they stood 'at the threshold of the most spectacular economic developments ever witnessed', as Joseph Schumpeter put it a century later: 'Vast possibilities matured into realities before their very eyes. Nevertheless, they saw nothing but cramped economies struggling with ever-decreasing success for their daily bread.' This was because their world view was dominated by the idea of diminishing returns. Ricardo, for example, watching local farmers struggle with bad harvests in the 1810s, agreed with his friend Malthus that corn yields must stagnate, because the best land was already in cultivation and every marginal acre brought under the plough would be worse than the one before. So Smith's division of labour, and Ricardo's comparative advantage, could improve the lot of people only up to a point. These were just a more efficient way of squeezing prosperity out of a limited system. Even after living standards began to rocket upwards in Britain from the 1830s, Mill saw it as a flash in the pan. Diminishing returns would soon set in. In the 1930s and 1940s, John Maynard Keynes and Alvin Hansen saw the Great Depression as evidence that some limit of human prosperity had been reached. Demand for cars and electricity was satiated and returns on capital were falling, so the world faced a future of chronic unemployment, once the sugar rush of war spending faded. The end of the Second World War would bring stagnation and misery. Again in the 1970s, and in the 2010s, there was widespread talk of sharing out the existing wealth of society rather than hoping living standards could go higher. Stagnationism has its fans in every generation.

Yet repeatedly the opposite happened. Far from diminishing, returns kept increasing thanks to mechanisation and the application of cheap energy. The productivity of a worker, rather than reaching a plateau, just kept on rising. The more steel was produced, the cheaper it got. The cheaper mobile phones grew, the more we used them. As Britain and then the world grew more populous, the more mouths there were to feed, the fewer people starved: famine is now largely unknown in a world of seven billion people, whereas it was a regular guest when there were two billion. Even Ricardo's wheat yields, from British fields that had been ploughed for millennia, began to accelerate upwards in the second half of the twentieth century thanks to fertilisers, pesticides and plant breeding. By the early twenty-first century, industrialisation had spread high living standards to almost every corner of the globe, in direct contradiction to the pessimistic fears of many that they would forever remain a Western privilege. China, a country mired in misery for centuries, and plunged into horror for decades, sprang to life and saw its billion people create the world's largest market.

What was happening? Nobody had set out to cause this phenomenon of global economic growth, or even predicted that it was possible. It simply emerged and spread as the nineteenth and twentieth centuries unrolled. It evolved.

All along, economists struggled to explain it, and they still do. Karl Marx had a stab at it, recognising the fact of industrial change, but he swallowed Ricardo's idea that mechanisation would leave an army of unemployed workers for the capitalist to exploit, whereas in the event both the number of jobs and the share of rewards going to workers went steadily upwards in industrial economies. The 'marginalist' revolution in economics, led by Carl Menger, Léon Walras and Stanley Jevons and culminating in the synthesis of Alfred Marshall, shifted the focus of price setting to the consumer rather than the producer, but left the question of increasing returns largely unanswered. In place of diminishing returns, they produced the idea of an equilibrium – a steady state of perfect competition towards which the economic system tended to move once information is easily available.

Then came Joseph Schumpeter, with his relentless focus on innovation and his insistence that there was no equilibrium, but an unfolding of incessant, dynamic change. In his *Theory of Economic Development*, written while he was at the University of Czernowitz in 1909, Schumpeter was the first economist to insist that the role of the entrepreneur was crucial. Far from being parasitic exploiters of the workers, most businessmen were innovators looking to outwit their rivals, by doing things better or cheaper, and in doing so they inevitably brought improvements to the living standards of consumers. Most of the so-called robber barons got rich by cutting the price of goods, not raising them. Innovation was the key consequence of free enterprise, dwarfing gains from trade, efficiencies of specialisation and improvements by practice. In a famous phrase introduced in his book *Capitalism, Socialism and Democracy* in 1942,

Schumpeter saw 'creative destruction' as the key to economic progress, and the 'essential fact about capitalism'. For new firms and technologies to emerge, old ones had to die. There is a 'perennial gale of creative destruction'. Or, as Nassim Taleb puts it, for the economy to be antifragile (strengthened by running risks), individual firms must be fragile. The restaurant business is robust and successful precisely because individual restaurants are vulnerable and short-lived. Taleb wishes that society honoured ruined entrepreneurs as richly as it honours fallen soldiers.

Schumpeter was explicitly biological in his reasoning, referring to economic change as a process of 'industrial mutation'. He saw that an economy is like an ecosystem, in which the struggle for existence causes businesses and products to compete and to change. He also saw that without risk-taking entrepreneurs, this economic evolution would not happen. Schumpeter's evolutionary perspective has recently been extended by the entrepreneur Nick Hanauer and the economist Eric Beinhocker. They argue that markets, like ecosystems, work not because they are efficient, but because they are effective, because they provide solutions to problems that face customers (or organisms). And the beauty of commerce is that when it works it rewards people for solving other people's problems. It is 'best understood as an evolutionary system, constantly creating and trying out new solutions to problems in a similar way to how evolution works in nature. Some solutions are "fitter" than others. The fittest survive and propagate. The unfit die.'

A corollary of this perspective is that there is no such thing as a perfect market, an equilibrium or an end state. Interestingly, ecologists have been coming gradually to the same conclusion as economists. They have started to move away from equilibrium thinking in recent years towards a much more dynamic view of ecosystems. Not only have they come to appreciate the way that climate changes, as ice ages wax and wane, they have even begun to realise that forests are in a state of continual change, as one species of tree succeeds another in a particular spot. There is no steady-state 'climax', just continual change. Not that the news has reached most policy-makers. The ecologist Daniel Botkin has complained that while ecologists agree that nature changes, when asked to design a policy, they nearly always come up with a 'balance-of-nature policy' that assumes an equilibrium. Call it the dynamic revolution in both economics and ecology.

Innovationism

Since Schumpeter, economists have taken up the challenge of explaining this innovation thing that has been happening to us and driving up living standards. Robert Solow in the 1950s was able to tease out just how much it was contributing by counting the contribution of capital and labour, and deducing that the rest (87.5 per cent) of the change in living standards must be due to technological change. It is technological change that is the chief source of increasing returns: of the fact that economic growth for the world as a whole shows no sign of reaching a plateau.

Little wonder, then, that Deirdre McCloskey describes the system that produced the great enrichment of the past two centuries as 'innovationism' rather than 'capitalism'. The new and crucial ingredient was not the availability of capital, but the advent of market-tested, consumer-driven innovation. She locates the cause of the Industrial Revolution in the decentralisation of the production and testing of new ideas: ordinary people were able to contribute to and to choose the products and services they preferred, which drove continuous innovation. For the Industrial Revolution to happen, trial and error had to become respectable. As she put it in a lecture in India in 2014, the enrichment of the poor has resulted not from charity, planning, protection, regulation or trade unions, all of which merely redistribute money, but from innovation caused by markets, which have not been bad for the poor: 'The sole reliable good for the poor, on the contrary, has been the liberating and the honoring of market-tested improvement and supply.'

But does innovation just happen, or is it in itself a product that can be created? This is the question that Paul Romer tackled in the 1990s with his theory of endogenous growth. Technical advances are not just by-products of growth, he argued, but investments that firms can deliberately make. Given the right institutions – a market in which to sell your product, the rule of law to prevent theft, a decent system of finance and taxation to incentivise you, some intellectual property protection, but not too much – you can set out to make an innovation and reap the rewards from it, despite sharing it with the world, in the same way you can set out to build a machine. That is roughly what the various firms around the world that are, as of this writing, offering mobile taxi-hiring services (Uber, Lyft, Hailo and so on) are doing – investing in innovation itself. But apart from some vague hand-waving about institutions, economists still have very little to offer in the way of prescriptions for innovation, except to say they know it will happen in open, free societies connected to the rest of the world by trade so that ideas can meet and mate.

And even these explanations come long after the phenomenon itself. A surge of innovation has lowered the cost of fulfilling people's needs and lowered the amount of time they had to work to fill those needs, thus raising living standards decade after decade, without anybody really being able to explain why and how it happened, let alone cause it to happen. Do you see why I am no fan of experts, policies and strategies? We were the unwitting guinea pigs in an enormous and global evolutionary surge, and it came from that most mysterious of human institutions – the market.

I suspect that we will never explain innovation fully, for the best of Lucretian reasons – that an explanation would require omniscience, the centralising of knowledge that is widely dispersed. Just as the Industrial Revolution took the world by surprise because it emerged from thousands of individual fragments of partial knowledge, rather than as a plan, so every innovation to this day is the result of thousands of people exchanging ideas. We can never predict innovation; we can only say that it will mysteriously emerge whenever people are free to exchange. The economist Larry Summers tells his students: 'Things will happen in well-organised efforts without direction, controls, plans. That's the consensus among economists.'

Adam Darwin

Smithism, like Darwinism, is a theory of evolutionary mechanism: a hypothesis as to what causes change that is not random, but not directed either. As I argued in a lecture in 2012, today few people appreciate just how similar the arguments made by Smith and Darwin are. Generally, Adam Smith is championed by the political right, Charles Darwin more often by the left. In, say, Texas, where Smith's emergent, decentralised economics is all the rage, Darwin is frequently reviled for his contradiction of *dirigiste* creation. In the average British university, by contrast, you will find fervent believers in the emergent, decentralised properties of genomes and ecosystems who yet demand *dirigiste* policy to bring order to the economy and society. But if life needs no intelligent designer, then why should the market need a central planner? Where Darwin defenestrated God, Smith just as surely defenestrated Leviathan. Society, he said, is a spontaneously ordered phenomenon. And Smith faces the same baffled incredulity – How can society work for the good of all without direction? – that Darwin faces.

Economic evolution is a process of variation and selection, just like biological evolution. Indeed, there is an even closer parallel. As I argued in *The Rational Optimist*, exchange plays the same vital role in economic evolution as sex plays in biological evolution. Without sex, natural selection is not a cumulative force. Mutations that occur in different lineages cannot come together; the struggle for existence must choose between them. Suppose, for example, that two different individuals in an ancestral mammal species invented fur and milk (two great mammalian innovations) around the same time. Had the species been an asexual one, reproducing by a form of cloning, then the two innovations would have remained in different, competing lineages. Natural selection would – in effect – have to choose which of the two it preferred. In a sexual species, however, an individual can inherit milk genes from its mother and fur genes from its father. Sex enables individuals to draw upon innovations that occur anywhere in the species.

Exchange has the same effect on economic evolution. In a society that is not open to trade, one tribe might invent the bow and arrow, while another invents fire. The two tribes will now compete, and if the one with fire prevails, the one with bows and arrows will die out, taking their idea with them. In a society that trades, the fire-makers can have bows and arrows, and vice versa. Trade makes innovation a cumulative phenomenon. Lack of trade may well be what held back the otherwise intelligent Neanderthals. It is certainly what held back many isolated human tribes in competition with those that could draw upon much wider sources of innovation. Instead of relying on your own village for innovation, you can get ideas from elsewhere. I make use of thousands of brilliant innovations every day. Very few of them were made in my own country, let alone my own village.

The mighty consumer

When it comes to economics, pretty well everybody is still in thrall to creationism. Don Boudreaux, an economist, thinks most people are secular theists who believe that social order is the result of 'some higher power that designs, intends, imposes, and guides willfully the order that we see about us'. They think that 'most of the economic and social order that we experience about us is the result of government and, hence, would necessarily disappear or collapse into disarray were government to disappear or fail to perform its duty well'.

You will often hear people say that free markets have been discredited, as they sip cups of coffee while sitting on chairs, wearing clothes and checking text messages – each of which was supplied by hundreds, thousands of producers whose beautifully coordinated collaboration was unplanned but achieved by 'market forces'. You will hear people say that none of this could happen without government to provide the roads, the traffic lights, the air-traffic control, the police, the law that make commerce possible. True, and Adam Smith was the first person to observe that it is the duty of the state to protect trade from pirates, predators and monopolists. He was no anarchist. But to leap from this to the conclusion that social order is consciously designed or enforced is absurd. Who decreed that coffee shops should take the form they do? The customers.

As Ludwig von Mises pointed out in 1944, the real bosses in a market economy are the consumers.

> They, by their buying and by their abstention from buying, decide who should own the capital and run the plants. They determine what should be produced and in what quantity and quality. Their attitudes result either in profit or in loss for the enterpriser. They make poor men rich and rich men poor. They are no easy bosses. They are full of whims and fancies, changeable and unpredictable. They do not care a whit for past merit. As soon as something is offered to them that they like better or is cheaper, they desert their old purveyors.

Look at the vulnerability of large companies when they do something their customer-bosses do not like. Coca-Cola's New Coke was an instant disaster from which the company beat a humiliating retreat. Large companies are vulnerable to the whims of their customers, and they know it. Free-market commerce is the only system of human organisation yet devised where ordinary people are in charge – unlike feudalism, communism, fascism, slavery and socialism.

It is axiomatic among right-thinking people that there are many things the market cannot provide, and therefore the state must. The sheer magical mysticism inherent in this thought is rarely examined. Because the market cannot do something, why must we assume that the state knows better how to do it? All too often this is, to borrow a phrase from Don Boudreaux, to 'assume a miracle'. Yet the history of government over the past few centuries is that when the state steps in to provide something that was underprovided by people for themselves, things do not necessarily improve; often they get worse. Market failure is a favourite phrase; government failure is not.

Take six basic needs of a human being: food, clothing, health, education, shelter and transport. Roughly speaking, in most countries the market provides food and clothing, the state provides healthcare and education, while shelter and transport are provided by a mixture of the two – private firms with semi-monopolistic privileges supplied by government: crony capitalism, in a phrase.

Is it not striking that the cost of food and clothing has gone steadily downwards over the past fifty years, while the cost of healthcare and education has gone steadily upwards? In 1969 the average American household spent 22 per cent of consumption expenditure on food and 8 per cent on clothing. Now it spends 13 per cent on food and 4 per cent on clothing. Yet the quality and diversity of both food and clothing have improved immeasurably since 1969. By contrast, the consumption of healthcare has more than doubled, from 9 per cent to 22 per cent of household expenditure, and the consumption of education has trebled, from 1 per cent to 3 per cent. The quality of both is the subject of frequent lament and complaint. Cost keeps going up, quality not so much, and innovation is sluggish. As for transport and shelter, broadly speaking the parts that the market supplies – budget airlines, house-building – have got cheaper and better, while the parts that the state supplies – infrastructure and land planning – have got more expensive and slower.

Prima facie, therefore, the market is doing a better job of supplying people with what they need (it is also pretty good at supplying what they desire, like entertainment). But perhaps the comparison is unfair. Healthcare is bound to inflate in cost because of new procedures and longer lifespans. There may be a similar excuse for education, but I cannot put my finger on it at the moment.

Besides, it is simply a fact of life that health and education must be supplied by the state, because – well, why? Because the market is not prepared to step forward? Hardly. Because the market would cheat the ill-informed consumer? It does not do so in the case of food and clothing, at least not much. Because the market would supply only the wealthy? Again, food and clothing suggest otherwise, as does the history of the medical profession. In times past doctors often charged their wealthy clients more than their poor ones, using the former to subsidise the treatment of the latter. Before Medicaid and Medicare existed, writes the American politician and former doctor Ron Paul, 'every physician understood that he or she had a responsibility towards the less fortunate, and free medical care for the poor was the norm'.

An alternative to Leviathan

Real counterfactuals are at hand to test the proposition that healthcare provision would be cheaper and better if consumers were in charge through the market, rather than government officials through the state; and that food provision would be more expensive and worse if the state were in charge, rather than the consumer. 'Were we directed from Washington when to sow and when to reap, we should soon want bread,' wrote Thomas Jefferson, presciently. In the Soviet Union the state had – and in today's North Korea it has – a monopoly in the provision of food, from field to fork. The result was (and in North Korea is) dismal productivity, frequent shortages, scandalous lapses in quality, and rationing by queue – and by privilege. These are exactly the features that have dominated Britain's healthcare debate over the past few years. In the provision of food, the consumer is the regulator enforcing better practice and lower cost; in the provision of healthcare, accountability, via government, is remote and slow, and regulators are frequently held captive by producers.

But the most startling counterfactual is the history of friendly societies. As the social scientist David Green has demonstrated, in Britain friendly societies grew like weeds in the late nineteenth and early twentieth centuries. By 1910 three-quarters of British manual workers were members. Friendly societies were small, local unions of workers which bought health insurance on behalf of their members and negotiated care from doctors and hospitals. Doctors who failed to do a good job were dropped, so they were directly accountable to their patients in a way that they simply are not today, when they answer to commissions and managers. Competition between them kept salaries modest, but doctors were still well paid. This was, therefore, a national health service that was widespread though not universal, growing rapidly, and reassuring to working people, since it ensured that they had access to more expensive treatments they could not afford to buy directly. It had emerged spontaneously and organically, and membership of the movement had doubled in fifteen years. This was socialism without the state. There is no doubt that it would have continued to expand and evolve.

But the friendly societies had their enemies. The commercial insurance companies, organised into a cartel called the Combine, felt threatened by these rival arrangements, and campaigned against them. So did the doctors' union, the British Medical Association, which in the writer Dominic Frisby's words 'loathed the fact that, under the Friendly Society system, the customer, or patient, had control and doctors were accountable to them'. Snooty doctors disliked being at the beck and call of societies of workers, let alone having to compete on price. These opponents successfully lobbied the Chancellor of the Exchequer, David Lloyd George, to introduce a system of 'national insurance', which was nothing of the sort, but was simply a poll tax deducted at source. Lloyd George used the proceeds of the tax to double the minimum pay of doctors, effectively transferring wealth from poor workers to rich doctors. With doctors' services more expensive, the entire friendly society system immediately began to wither. In 1948 the healthcare industry was nationalised and the state began to provide all medical care, free at the point of delivery and decided for you by he who knows best in government.

Now, of course, there are good doctors in the nationalised system, and there were bad ones in the friendly society system; and of course healthcare has changed drastically since the days of the friendly societies, thanks to science and technology. But the system would have evolved, innovated, kept pace with the growth of wages and encouraged discovery. We can never know what a twenty-first-century friendly society health system would have looked like, but everything we know about the evolution of market-driven systems suggests that it would have catered to the needs of all, especially the poor, in a way that would have progressed very rapidly. It would be as different today as a supermarket is from a corner shop in 1910.

The worst of it is that the British National Health Service is not fully nationalised at all. The provision of care is nationalised, and decided for you by committees. But the workers who treat you, the doctors, are private contractors with generous terms. As with so much of modern life, the state has socialised the cost and privatised the reward. That is what tax-funded monarchs did, what tithe-funded monks did, what prize-seeking naval captains did, what corrupt colonial nabobs did – and it is what today's broadcasters, artists, scientists, civil servants and doctors do, almost to a man and woman. They rely heavily on the state for their wages, budgets or grants. This is the modern clerisy.

Around them crowd thousands more whose income is privately earned in fees, but to a startling degree comes straight from the lavish coffers of the state: bankers, lawyers, architects, environmentalists and others. The affairs of Parliament, I am (not entirely) surprised to discover, are dominated by rent-seeking professions demanding that Leviathan sluices money towards them, whether to administer regulations, inquire into trends, judge cases, or build power stations. Businessmen are the worst. It is a myth that they love the evolutionary, free market; in practice they seek privilege and monopoly at the drop of a legislative hat. Adam Smith was not wrong when he said, 'People of the same trade seldom meet together, even for merriment and diversion, but the conversation ends in a conspiracy against the public, or in some contrivance to raise prices.'

7

The Evolution of Technology

For then, since gold was soft and blunted easily, men would deem
It useless, but bronze was a metal held in high esteem.
Now the opposite: bronze is held cheap, while gold is prime.
And so the seasons of all things roll with the round of time:
What once was valuable, at length is held of no account,
While yet the worth of that which was despised begins to mount.

Lucretius, *De Rerum Natura*, Book 5, lines 1272–7

The light bulb is both a metaphor for invention and in itself a neat invention. Imagine: you have to think of the idea of making a filament glow incandescently (but not burn up) when an electric current is put through it. You have to encase the thing in glass and then you have to pump the air out to make a partial vacuum. Not a straightforward idea by any means. As inventions go, it's probably done more good and less harm than almost any other idea. It brought cheap light to night and to winter for billions of people; it banished the smoke and fire risk of candles and kerosene; it enabled education to reach far more children. As I mentioned in a previous book, it brought down the time you need to work on the average wage to earn an hour of artificial light to less than a second, compared with minutes for kerosene lamps in their heyday, and hours for tallow candles. Sure, it's been used in interrogations, but let's stay positive and thank God for Thomas Edison.

Suppose Thomas Edison had died of an electric shock before thinking up the light bulb. Would history have been radically different? Of course not. Somebody else would have come up with the idea. Others did. Where I live, we tend to call the Newcastle hero Joseph Swan the inventor of the incandescent bulb, and we are not wrong. He demonstrated his version slightly before Edison, and they settled their dispute by forming a joint company. In Russia, they credit Alexander Lodygin. In fact there are no fewer than twenty-three people who deserve the credit for inventing some version of the incandescent bulb before Edison, according to a history of the invention written by Robert Friedel, Paul Israel and Bernard Finn. Though it may not seem obvious to many of us, it was utterly inevitable once electricity became commonplace that light bulbs would be invented when they were. For all his brilliance, Edison was wholly dispensable and unnecessary. Consider the fact that Elisha Gray and Alexander Graham Bell filed for a patent on the telephone on the very same day. If one of them had been trampled by a horse en route to the patent office, history would have been much the same.

I am going to argue that invention is an evolutionary phenomenon. The way I was taught, technology was invented by god-like geniuses who stumbled upon ideas that changed the world. The steam engine, light bulb, jet engine, atom bomb, transistor – they came about because of Stephenson, Edison, Whittle, Oppenheimer, Shockley. These were the creators. We not only credit inventors with changing the world; we shower them with prizes and patents.

But do they really deserve it? Grateful as I am to Sergey Brin for the search engine, and to Steve Jobs for my Macbook, and to Brahmagupta (via Al Khwarizmi and Fibonacci) for zero, do I really think that if they had not been born, the search engine, the user-friendly laptop and zero would not by now exist? Just as the light bulb was 'ripe' for discovery in 1870, so the search engine was 'ripe' for discovery in 1990. By the time Google came along in 1996, there were already lots of search engines: Archie, Veronica, Excite, Infoseek, Altavista, Galaxy, Webcrawler, Yahoo, Lycos, Looksmart . . . to name just the most prominent. Perhaps none was at the time as good as Google, but they would have got better.

The truth is, almost all discoveries and inventions occur to different people simultaneously, and result in furious disputes between rivals who accuse each other of intellectual theft. In the early days of electricity, Park Benjamin, author of *The Age of Electricity*, observed that 'not an electrical invention of any importance has been made but that the honour of its origin has been claimed by more than one person'.

This phenomenon is so common that it must be telling us something about the inevitability of invention. As Kevin Kelly documents in his book *What Technology Wants*, we know of six different inventors of the thermometer, three of the hypodermic needle, four of vaccination, four of decimal fractions, five of the electric telegraph, four of photography, three of logarithms, five of the steamboat, six of the electric railroad. This is either redundancy on a grand scale, or a mighty coincidence. It was inevitable that these things would be invented or discovered just about when they were. The history of inventions, writes the historian Alfred Kroeber, is 'one endless chain of parallel instances'.

It's just as true in science as in technology. Boyle's Law in English-speaking countries is the same thing as Mariotte's Law in French-speaking countries. Isaac Newton vented paroxysms of fury at Gottfried Leibniz for claiming, correctly, to have invented the calculus independently. Charles Darwin was prodded into publishing his theory at last by Alfred Wallace having precisely the same idea, after reading precisely the same book (Malthus's *Essay on Population*). Britain and France almost went to war in the 1840s when the dispute between John Adams and Urbain Le Verrier over who discovered Neptune reached fever pitch in the press: they both found the planet. The tumour-suppressor gene p53, the disabling of which is crucial to the malignancy of most cancers, was discovered independently in 1979 in four different laboratories in London, Paris, New Jersey and New York.

Not even Einstein escapes the demolition of the unique discoverer. The ideas he put together as special relativity in 1905 were already beginning to be thought by others, notably Henri Poincaré and Hendrik Lorentz. This is not to diminish Einstein's ability. Clearly, he got there quicker and more deeply than anybody else. But it is impossible to imagine relativity remaining undiscovered for long in the first half of the twentieth century, just as it is impossible to imagine the genetic code remaining undiscovered for long in the second half. The discovery of the double helix in 1953 remains plagued to this day by accusations that too much credit went to the first two people to solve the structure, and not to those who did some of the hard work that led to the insight. As Francis Crick pondered of his partner in the elucidation of the double helix, James Watson: 'If Jim had been killed by a tennis ball, I am reasonably sure I would not have solved the structure alone, but who would?' There were plenty of candidates: Maurice Wilkins, Rosalind Franklin, Raymond Gosling, Linus Pauling, Sven Furberg, and others. The double helix and the genetic code would not have remained hidden for long.

Gregor Mendel, the father of genetics, is an interesting exception to the rule of simultaneous discovery. His revelation of independently assorting, apparently indivisible particles of inheritance (genes) stood alone in the 1860s – though you can make a case that a chap called Thomas Knight had glimpsed the insight a few decades before, when he noticed that violet-flowered peas crossed with white-flowered peas produced mainly violet-flowered offspring. But interestingly, Mendel, just as much as Knight, was before his time. The idea was still unripe, and since it fitted neither the preconceptions nor the needs of scientists, it was ignored, indeed forgotten. Quite suddenly in 1900, thirty-five years afterwards, three different scientists stumbled upon the same genetic insights and belatedly – after some prodding – gave the credit to Mendel. A case of simultaneous rediscovery. The point is that genetics was ready to begin in 1900, but not in 1865. Just as you cannot stop discovery happening, perhaps you can't hurry it much either.

In case you still think there is a lingering smell of plagiarism in the many episodes of simultaneous discovery, consider the idea of a nuclear chain reaction. There is a formula, called the four-factor formula, that allows you to calculate the critical mass necessary to cause a chain reaction. Working entirely in secret, six different teams found the four-factor formula, in America (thrice), in France, Germany and the Soviet Union. The Japanese also came close, and the British contributed to the American efforts.

Inexorable technological progress

Simultaneous discovery and invention mean that both patents and Nobel Prizes are fundamentally unfair things. And indeed, it is rare for a Nobel Prize not to leave in its wake a train of bitterly disappointed individuals with very good cause to be bitterly disappointed. Nor is this matter confined to science and technology. Kevin Kelly catalogues many instances of simultaneous release of films with similar plots and of books with similar themes. As he drily remarks, after listing the many uncanny premonitions of Harry Potter themes in obscure books that J.K. Rowling never read: 'Because a lot of money swirls around Harry Potter we have discovered that, strange as it sounds, stories of boy wizards in magical schools with pet owls who enter their other worlds through railway station platforms are inevitable at this point in Western culture.'

Two other phenomena underline the overwhelming inevitability in the progress of technology. The first is the equivalent of what biologists call convergent evolution – the appearance of the same solution to a particular problem in widely different places. Thus ancient Egyptians and ancient Australians both invented curved boomerangs without conferring. Amazonian and Bornean hunter-gatherers both invented blowguns to fire poisoned darts at monkeys and birds. Remarkably, both lit upon the counterintuitive idea that to use them accurately requires holding them with both hands close to the face and turning them in slow circles rather than trying to keep them perfectly still.

The other hint as to the inevitability of technological change comes from the way that progress happens incrementally and inexorably – and is impossible to prevent. The clearest example of this is Moore's Law. In 1965, the computer expert Gordon Moore drew a little graph of the number of 'components per integrated function' on a silicon chip against time. On the basis of just five data points, he deduced that the number of transistors on a chip seemed to be doubling every year and a half. He consulted a friend and colleague, Carver Mead, who did some calculations to try to work out what the limits of this shrinkage would be. It was Mead who spotted that the shrinkage was not just making chips denser, it was making them more efficient. Speed goes up, power consumption goes down, system reliability improves, and cost falls. In Moore's words: 'By making things smaller, everything gets better simultaneously. There is little need for trade-offs.'

Eerily, the progress of the computer has followed Moore's Law ever since, with extraordinarily little deviation. Moore himself expected it to hit a limit when the size of each transistor reached 250 nanometres in diameter, but it passed that point in 1997 and kept on plunging. What explains this extraordinary, predictable regularity? You might say, well, it is a self-fulfilling prophecy, that it's because technologists know that improvement can be done, so they make sure it is done at that rate. But surely an entrepreneur who told his people to leapfrog ahead would gain a great advantage? Yet this never seems to happen. It was not possible to imagine, let alone build, the computer of 2015 in

2005, let alone in 1965 – the intermediate steps were crucial. As in the evolution of living species, each intervening step must be a viable organism.

That does not stop clever people using Moore's Law as a guide to the future. When Alvy Ray Smith and Ed Catmull founded Pixar to make computer-animated movies, it was only after twice postponing any project of the kind because they knew computing was still too slow and costly. After the second abortive attempt Smith predicted, using Moore's Law, that it would be five years before computer-animated movies would prove viable, because Moore's Law can be restated as 'Computers improve by an order of magnitude every five years.' So when Disney approached Pixar five years later to make *Toy Story*, they said yes, and the rest is history.

A few years ago, Ray Kurzweil made a startling discovery: that Moore's Law was being obeyed before silicon chips even existed. By extrapolating the power of computers back to the early twentieth century, when they used different technologies altogether, he drew a straight line on a logarithmic curve. Before the integrated circuit even existed, the electromechanical relay, vacuum tube and the transistor had all improved along the very same trajectory. To put it another way, the amount of computing power that you can buy for £100 has doubled every two years for a century. And if Moore's Law has continued through technological change-overs, then there's no reason to think it will not happen again. When chips eventually reach their miniaturisation limit, the plummeting cost will continue in another technology.

Nor is Moore's Law the only such regularity to emerge in the computer age. Kryder's Law says that the cost per performance of hard disk computer storage is rising exponentially, at 40 per cent a year. Cooper's Law finds that the number of possible simultaneous wireless communications has doubled every thirty months since 1895, when Marconi first broadcast. These are largely independent of Moore's Law. And spookily, these laws have marched implacably onwards through the upheavals of the twentieth century without breaking step. As I put it in an article for the *Wall Street Journal*: 'How is it possible that the Great Depression did not slow down technological progress? Why didn't the great infusion of technology spending during World War II accelerate it?'

The explanation for the bizarre regularity of Moore's Law and its brethren seems to be that technology is driving its own progress. Each technology is necessary for the next technology. One of those who makes Moore's Law happen describes his role thus: 'We implement each step to see if it actually works, then gain the courage, the insight and the engineering mastery to proceed to the next step.'

And indeed this is the story of technology, from the Stone Age to the present day, on all continents: wherever you look, technology proceeds in a stately way from each tool to the next, rarely leapfrogging or sidestepping. As Kelly remarks, the sequence is always uniform, and is significantly correlated on different continents: 'Knifepoints always follow fire, human burials always follow knifepoints, and the arch precedes welding.' To this day, it is very hard for a country to become a knowledge economy without being an agricultural success and then a manufacturing success first. That's the path Japan, South Korea, China, India, Mauritius and Brazil have followed in recent years, and it's the path that Britain and America followed at a more leisurely pace in the eighteenth, nineteenth and twentieth centuries.

This path dependence is obvious in some ways. There's not much point in mining uranium till you have invented steel, cement, electricity and computing, and understood nuclear physics. Technology proceeds, like evolution, to the 'adjacent possible', a phrase coined by the evolutionary biologist Stuart Kauffman. It does not leap far into the future. I recently tried to think of examples of inventions that came long after their time, that should have been invented much sooner than they were – things we take for granted now and that would have been great for our grandparents to have had. It's surprisingly hard to come up with them. I thought wheeled suitcases were a good example, recalling all those days when I lugged a heavy bag to a railway station in my youth. Bernard Sadow patented the wheeled suitcase in 1970, after watching a porter wheeling bags on a cart in an airport. His patent was for four-wheeled cases dragged by a lead like a dog. Lots of suitcase manufacturers refused to take him seriously. Seventeen years later – in 1987! – Robert Plath, an airline pilot, came up with the idea of the two-wheeled bag with the telescopic handle. Surely both of these could have come much earlier? Actually, I am not so sure. Before 1970 airports were smaller, you could drive right up to them, check-in was close, there were porters to help at railway stations with wheels on their carts – so why go to the trouble of installing wheels inside the suitcase, especially when they would have to be made of heavy steel? In retrospect, 1970 was probably the moment when plastic and aluminium made carrying wheels with the case practical for the first time. In practice, inventions rarely run late. They turn up at just the moment in history when it makes most sense that they do so. The first laptop, in 1982, came when computers had at last got small enough not to crush your knees through the floor.

The sea fashions boats

Kevin Kelly's 2010 book is not the only one in recent years that has begun to describe technology in evolutionary terms. In 2009 Brian Arthur of the Santa Fe Institute published a book called *The Nature of Technology: What it is and How it Evolves,* in which he concluded 'that novel technologies arise by combination of existing technologies and that (therefore) existing technologies beget further technologies . . . we can say that technology creates itself out of itself'. He saw explicitly Darwinian themes in the steady accumulation of beneficial innovations within the progress of technology. I made a similar point in my 2010 book *The*

Rational Optimist: How Prosperity Evolves, in drawing attention to the similarity between the recombination of genes as a result of sex to produce biological novelty and the recombination of ideas as a result of trade to produce technological novelty: 'ideas having sex' explains why innovation has tended to happen in open societies indulging in enthusiastic free trade. The same year, Steven Berlin Johnson published *Where Good Ideas Come From: The Natural History of Innovation*, and developed the idea that the story of technology, like biological evolution, is a 'gradual but relentless probing of the adjacent possible, each new innovation opening up new paths to explore'. The economics writer Tim Harford, in his 2011 book *Adapt: Why Success Always Starts With Failure*, pointed out that 'trial and error is a tremendously powerful process for solving problems in a complex world, while expert leadership is not'. Intelligent design is just as bad at explaining society as it is at explaining evolution.

Either we five authors all plagiarised each other, or around the end of the first decade of the twenty-first century there was a simultaneous discovery (ha!) of detailed evolutionary parallels in the story of technology. The idea was ripe. Of course we were not the first to discern 'Darwin among the machines' – that was the title of an essay of 1863 by Samuel Butler. A little later the anthropologist Augustus Pitt-Rivers drew up family trees of aboriginal weapons, demonstrating descent with modification, a diagnostic feature of evolution.

From these stirrings emerged the first challenge to the heroic view of invention as the fortuitous eurekas of men of genius. Instead, the incremental but inexorable progress of technology began to emerge. In the 1920s the American sociologist Colum Gilfillan traced the pedigree of ships from dugout canoes to steamships, implying that there was a gradualism about the progress of technology that stories of sudden invention disguised, and an inevitability about each step once the previous one had been taken. In 1922 William Ogburn developed a fully-fledged theory of emergent invention, arguing that 'the more there is to invent with, the greater will be the number of inventions'. The economists Joseph Schumpeter and Friedrich Hayek both saw the economy in explicitly Darwinian ways: as a system where ideas recombine and trends emerge rather than are imposed. In 1988 George Basalla wrote a book called *The Evolution of Technology*, which stressed the continuity of successive innovations. He pointed out that Eli Whitney's cotton gin was not conjured out of thin air, but adapted from the Indian charka or roller gins already in use. Basalla concluded that even such discrete jumps in technological history as the replacement of the propeller by the turbojet, or the vacuum diode by the transistor, are misleading. Both the turbojet and the transistor had long and gradual histories behind them, albeit in other applications – the turbine and the crystal radio receiver. Emphasising continuity, he pointed out that in their mechanism and design, the first motor cars were little more than four-wheeled bicycles with engines.

One of the most beautiful evolutionary insights about technology came in 1908 from the philosopher 'Alain' (real name Emile Chartier), who wrote of fishermen's boats:

> Every boat is copied from another boat . . . Let's reason as follows in the manner of Darwin. It is clear that a very badly made boat will end up at the bottom after one or two voyages and thus never be copied . . . One could then say, with complete rigor, that it is the sea herself who fashions the boats, choosing those which function and destroying the others.

It is the sea herself who fashions the boats. It's in this radical re-imagining that the new wave of thinking about the evolution of technology in the current century is turning the world upside down.

Much the same can be said of the market. Indeed, as Peter Drucker wrote in his classic 1954 business book *The Practice of Management*, customers shape companies in much the same way: 'It is the customer who determines what a business is. For it is the customer and he alone, who through being willing to pay for a good or for a service, converts economic resources into wealth, things into goods.'

The similarities between technology and biology are not confined to the observation that both show descent with modification and both evolve by trial and error. Biology and technology in the end boil down to systems of information. Just as a human body is the expression of information written in its DNA, and the fact that it is non-randomly arranged is an expression of 'information' – the opposite of entropy – so a steam engine, a light bulb or a software package is itself an ordered piece of information. Technology is in that sense a continuation of biological evolution – an imposition of informational order on a random world.

Moreover, increasingly, technology is developing the kind of autonomy that hitherto characterised biological entities. Brian Arthur argues that since technology is self-organising and can in effect reproduce, respond and adapt to its environment, while taking in and giving out energy to maintain its being, then it qualifies as a living organism, at least in the sense that a coral reef is a living thing. Sure, it could not exist without animals (people) to build and maintain it, but then that's true of a coral reef, too. And who knows when this will no longer be true of technology, and it will build and maintain itself? To Kevin Kelly, the 'technium' – Kelly's name for the evolving organism that our collective machinery comprises – is already 'a very complex organism that often follows its own urges'. It 'wants what every living system wants: to perpetuate itself'. By 2010 the internet had roughly as many hyperlinks as the brain has synapses, and a significant proportion of the whispering that goes on within the internet originates in devices rather than people. It is already virtually impossible to turn the internet off.

If it is true that the technium has its own evolutionary

momentum, then the way to develop new products is to encourage technological evolution rather than to try to design
new products. The aircraft manufacturer Lockheed had this
idea in the 1940s, when it set aside what it called the 'skunk
works', as a laboratory charged with playing with new designs
almost at random. It was from the skunk works that the U-2
and Blackbird reconnaissance planes and the Stealth bomber
emerged. Google has likewise turned itself into a trial-and-error
company, by encouraging employees to spend 20 per cent of their
time on their own projects. Some years ago the multinational
corporation Procter & Gamble turned its back on the idea of
proprietary and secretive research and went instead for 'open
innovation', in which it was prepared to bring in ideas from outside the firm by partnering with their creators. The project is
called 'Connect and Develop', and the company says it is bearing
fruit. For example, it created the Live Well Collaborative with
Cincinnati University and other partners to harvest ideas about
how to design products to meet the needs of the elderly. More
than twenty products came out of the initiative.

The implications of this new way of seeing technology, as an
autonomous, evolving entity that continues to progress whoever
is in charge, are startling. People are pawns in a process. We ride
rather than drive the innovation wave. Technology will find its
inventors, rather than vice versa. Short of bumping off half the
population, there's little we can do to stop it happening, and
even that might not work. Indeed, the history of technological
prohibitions is revealing. The Ming Chinese prohibited large
ships, the Shogun Japanese firearms, the medieval Italians silk-
spinning, the 1920s Americans alcohol. Such prohibitions can
last a long time – three centuries in the case of the Chinese and
Japanese examples – but eventually they come to an end, so long
as there is competition. Meanwhile, elsewhere in the world, these
technologies continued to develop.

Today it is impossible to imagine software development
coming to a halt. Somewhere in the world a nation will harbour

programmers, however strongly (say) the United Nations tries to enforce a ban on software development. (The idea is absurd, which makes my point.) It is easier to prohibit technological development in larger-scale technologies that require big investments and national regulations. So, for example, Europe has fairly successfully maintained a *de facto* ban on genetic modification of crops for two decades in the name of the precautionary principle, and it looks as if it may do the same for shale gas, thanks largely to the unpleasant sound of the word 'fracking'. But even here there is no hope of stopping these technologies globally. Genetic modification and fracking are thriving elsewhere, bringing down pesticide usage and carbon dioxide emissions respectively.

And if there is no stopping technology, perhaps there is no steering it either. In Kelly's words, 'the technium wants what evolution began'. Technological change is a far more spontaneous phenomenon than we realise. Out with the heroic, revolutionary story of the inventor, in with the inexorable, incremental, inevitable creep of innovation.

Patent scepticism

It will come as no surprise that, having argued for the incremental, inevitable and collective nature of innovation, I am not a fan of patents and copyright laws. They grant too much credit and reward to individuals, and imply that technology evolves by jerks. I am not convinced that they have played a vital role in encouraging creativity in Western societies, as is often claimed. Shakespeare wrote astonishing plays with no copyright protection: cheap copies, scribbled down by members of the audience, were hawked around London within weeks of performances.

The original idea of a patent, remember, was not to reward inventors with monopoly profits, but to encourage them to share their inventions. A certain amount of intellectual property law is plainly necessary to achieve this. But it has gone too far.

Most patents are now as much about defending monopoly and deterring rivals as about sharing ideas. And that discourages innovation. Many firms use patents as barriers to entry, suing upstart innovators who trespass on their intellectual property even en route to some other goal. In the years before World War I, aircraft-makers tied each other up in patent lawsuits and slowed down innovation until the US government stepped in. Much the same has happened with smartphones and biotechnology today. New entrants have to fight their way through 'patent thickets' if they are to build on existing technologies to make new ones. (I just breached copyright: the last four sentences are lifted straight from an article I wrote for the *Wall Street Journal*.)

Nor is it clear how patents are supposed to cope with simultaneous discovery. As I have pointed out, parallel invention is the rule, not the exception; yet the patent courts insist that somebody somewhere deserves priority and profit. The economist Alex Tabarrok drew a convex, curved graph to illustrate the point that a little intellectual property is better than none, but a lot is a bad thing. He thinks US patent law is well beyond the optimal point. He argued in his 2011 book *Launching the Innovation Renaissance* that in practice imitation is often more costly than innovation. So there is little need for intellectual property protection, because the learning curve of the imitator is so steep. Even if you had been free to copy Google's search engine in the late 1990s, by the time you had worked through all the hidden obstacles that Google had also worked through, you would have been years behind.

Copying is not cheap

As this illustrates, the chief reason that copying is not much cheaper than original discovery is 'tacit knowledge'. Most of the little tricks and short cuts that industrialists follow to achieve their results remain in their heads. Even the most explicit paper or patent application fails to reveal nearly enough to help another to retrace your steps through the maze of possible experiments. One study of lasers found that blueprints and written reports were quite inadequate to help others copy laser design: you had to go and talk to people who had done it. Friedrich Hayek made this point when he argued that: 'Knowledge of the circumstances of which we must make use never exists in concentrated or inte- grated form but solely as the dispersed bits of incomplete and frequently contradictory knowledge which all the separate indivi- duals possess.' Or as Michael Polanyi put it more succinctly: 'We can know more than we can tell.' Edwin Mansfield of the University of Pennsylvania studied the development of forty- eight chemical, pharmaceutical, electronic and machine goods in New England in the 1970s, and found that on average it cost 65 per cent as much money, and 70 per cent as much time, to copy as to invent the products. And this was amongst specialists with technical expertise. Copying from scratch would cost even more. Commercial companies do basic research because they know it enables them to acquire the tacit knowledge that leads to inno- vation.

The obvious exception to the rule that copying is expensive is pharmaceuticals, where imitation –'generics' – is clearly cheaper than innovation. This is largely a consequence of safety regu- lation by governments. The state's not unreasonable demand that new drugs prove in huge clinical trials that they are harmless and effective means that it costs billions to bring them to market. Plainly, after requiring drug companies to spend such vast sums, the government will have to grant them some monopoly once it has licensed the new pill. Yet even here there is plenty of evi- dence that Big Pharma spends much of its monopoly profit on marketing rather than discovery.

Science is the daughter of technology

Politicians believe that innovation can be turned on and off like a tap. It starts, you see, with pure scientific insights, which then get translated into applied science, which in turn become useful technology. So what you must do, as a patriotic legislator, is ensure there is a ready supply of money to scientists on the top floor of their ivory towers, and lo and behold, technology will come clanking out of the pipe at the bottom of the tower.

This 'linear model' of how science drives innovation and prosperity goes right back to Francis Bacon, the Jacobean Lord Chancellor who urged England to catch up with the Portuguese in their use of science to drive discovery and commercial gain. Supposedly Prince Henry the Navigator in the fifteenth century had invested heavily in map-making, nautical skills and navigation at a special school at his villa on Portugal's Sagres peninsula, which resulted in the exploration of Africa and great gains from trade. That's what Bacon wanted to copy.

> The West Indies had never been discovered if the use of the
> mariner's needle had not been first discovered . . . There is not
> any part of good government more worthy than the further
> endowment of the world with sound and fruitful knowledge.

Yet recent scholarship has exposed this tale as a myth, or rather a piece of Prince Henry's propaganda. Like most innovation, Portugal's navigational advances came about by trial and error among sailors, not by speculation among astronomers and cartographers. If anything, the scientists were driven by the needs of the explorers rather than the other way around.

Professor Terence Kealey, a biochemist turned economist, tells this story to illustrate how he believes the linear dogma so prevalent in the world of science and politics – that science drives innovation, which drives commerce – is mostly wrong. It misunderstands where innovation comes from. Indeed, it generally gets it backwards. Again and again, once you examine the history of innovation, you find scientific breakthroughs as the effect, not the cause, of technological change. It is no accident that astronomy blossomed in the wake of the age of exploration.

The steam engine owed almost nothing to the science of thermo-dynamics, but the science of thermodynamics owed almost every-thing to the steam engine. The flowering of chemistry in the late nineteenth and early twentieth centuries was driven by the needs of dye-makers. The discovery of the structure of DNA depended heavily on X-ray crystallography of biological molecules, a tech-nique developed in the wool industry to try to improve textiles.

And so on, through case after case. The mechanisation of the textile industry was at the very heart of the Industrial Revolution, with its jennies, frames, mules, flying shuttles and mills going down in history as milestones in the industrialisation of Lanca-shire and Yorkshire, leading to Britain's sudden enrichment and power. Yet nowhere among the journeymen and entrepreneurs who drove these changes can you find even a hint of science. Much the same is true of mobile telephony in the late twentieth century. You will search in vain for major contributions from universities to the cellphone revolution. In both cases, techno-logical advances were driven by practical men who tinkered till they had better machines; philosophical rumination was the last thing they did.

As Nassim Taleb insists, from the methods used by thirteenth-century architects building cathedrals to the development of modern computing, the story of technology is a story of rules of thumb, learning by apprenticeship, chance discoveries, trial and error, tinkering – what the French call 'bricolage'.

Technology comes from technology far more often than from science. And science comes from technology too. Of course, science may from time to time return the favour to technology. Biotechnology would not have been possible without the science of molecular biology, for example. But the Baconian model, with its one-way flow from science to technology, from philosophy to practice, is nonsense. There's a much stronger flow the other way: new technologies give academics things to study.

An example: in recent years it has become fashionable to argue that the hydraulic fracturing technology that made the shale-gas revolution possible originated in government-sponsored research, and was handed on a plate to industry. A report by California's Breakthrough Institute noted that microseismic imaging was developed by the federal Sandia National Laboratory, and 'proved absolutely essential for drillers to navigate and site their boreholes', which led Nick Steinsberger, an engineer at Mitchell Energy, to develop the technique called 'slickwater fracking'.

To find out if this was true, I spoke to one of hydraulic fracturing's principal pioneers, Chris Wright, whose company Pinnacle Technologies reinvented fracking in the late 1990s in a way that unlocked the vast gas resources in the Barnett shale, in and around Forth Worth, Texas. Utilised by George Mitchell, who was pursuing a long and determined obsession with getting the gas to flow out of the Barnett shale to which he had rights, Pinnacle's recipe – slick water rather than thick gel, under just the right pressure and with sand to prop open the fractures through multi-stage fracturing – proved revolutionary. It was seeing a presentation by Wright that persuaded Mitchell's Steinsberger to try slickwater fracking. But where did Pinnacle get the idea? Wright had hired Norm Wapinski from Sandia, a federal laboratory. But who had funded Wapinksi to work on the project at Sandia? The Gas Research Institute, an entirely privately funded gas-industry research coalition, whose money came from a voluntary levy on interstate gas pipelines. So the only federal involvement was to provide a space in which to work. As Wright comments: 'If I had not hired Norm from Sandia there would have been no government involvement.' This was just the start. Fracking still took many years and huge sums of money to bring to fruition as a workable technology. Most of that was done by industry. Government laboratories beat a path to Wright's door once he had begun to crack the problem, offering their services and their public money to his efforts to improve fracking still further, and to study just how fractures propagate in rocks a mile beneath the surface. They climbed on the bandwagon, and got some science to do as a result of the technology developed in industry – as they should. But government was not the well-spring.

As Adam Smith, looking around the factories of eighteenth-century Scotland, reported in *The Wealth of Nations*: 'a great part of the machines made use in manufactures . . . were originally the inventions of common workmen', and many improvements had been made 'by the ingenuity of the makers of the machines'. Smith dismissed universities even as a source of advances in philosophy. I am sorry to say this to my friends in academic ivory towers, whose work I greatly value, but if you think your cogitations are the source of most practical innovation, you are badly mistaken.

Science as a private good

It follows that there is less need for government to fund science: industry will do this itself. Having made innovations, it will then pay for research into the principles behind them, as it did with microseismic imaging and fracking. Having invented the steam engine, it will pay for thermodynamics. This conclusion of Terence Kealey's is so heretical as to be incomprehensible to most economists, as well as scientists. It has been an article of faith for decades in both of their professions that science would not get funded if government did not do it, and economic growth would not happen if science did not get funded by the taxpayer. This received wisdom has been handed down for more than half a century. It was the economist Robert Solow who demonstrated in 1957 that innovation in technology was the source of most economic growth – at least in societies that were not expanding their territory or growing their populations. It was his economist colleagues Richard Nelson and Kenneth Arrow who explained in 1959 and 1962 respectively that government funding of science was necessary, because it is cheaper to copy others than to do original research. This makes science a public good, a service, like the light from a lighthouse, that must be provided at public expense, because nobody will supply it for free. No private individual will do basic science, for the insights that follow from it will be freely available to his rivals.

'The problem with the papers of Nelson and Arrow,' writes Kealey, 'was that they were theoretical and one or two troublesome souls, on peering out of their economists' eyries, noted that in the real world there did seem to be some privately funded research happening.' Kealey argues that there is still no empirical demonstration of the need for public funding of research, and that the historical record suggests the opposite. In the late nineteenth and early twentieth centuries, Britain and the United States made huge contributions to science with negligible public funding, while Germany and France, with hefty public funding, achieved no greater results either in science or in economics. 'The industrialised nations whose governments invested least in science did best economically,' says Kealey, 'and they didn't do so badly in science either.'

To most people, the argument for public funding of science rests on a list of the discoveries made with public funds, from the internet (defence science in the United States) to the Higgs boson (particle physics at CERN in Switzerland). But that's highly misleading. Given that government has funded science munificently, it would be odd if it had not found out something. We learn nothing about what would have been discovered by alternative funding arrangements. And we can never know what discoveries were not made, because government funding of science inevitably crowded out much of the philanthropic and commercial funding, which might have had different priorities.

After World War II, Britain and the United States changed tack and began to fund science heavily from the public purse. With the success of war science and of Soviet state funding that led to Sputnik, it seemed obvious that state funding must make a difference. The true lesson – that Sputnik relied heavily on Robert Goddard's work, which had been funded by the Guggenheims – could have gone the other way. Yet there was no growth dividend for Britain and America from this science-funding rush. Their economies grew no faster than they had before.

In 2003, the OECD published a paper on 'sources of growth in OECD countries' between 1971 and 1998, finding to its explicit surprise that whereas privately funded research and development stimulated economic growth, publicly funded research had no economic impact whatsoever. None. This earth-shaking result has never been challenged or debunked. Yet it is so inconvenient to the argument that science needs public funding that it is ignored.

In 2007, Leo Sveikauskas of the Bureau of Economic Analysis concluded that returns from many forms of publicly financed R&D are near zero, and that 'many elements of university and government research have very low returns, overwhelmingly contribute to economic growth only indirectly, if at all'. As Walter Park of the American University concluded, the explanation of this discrepancy is that public funding of research almost certainly crowds out private funding. That is to say, if the government spends money on the wrong kind of science, it tends to stop people working on the right kind of science. But, given that the government takes more than one-third of a nation's GDP in most countries and spends it on something, it would be a pity if none of that money found its way to science, which is after all one of the great triumphs of our culture.

Innovation, then, is an emergent phenomenon. The policies that have been tried to get it going – patents, prizes, government funding of science – may sometimes help, but are generally splendidly unpredictable. Where conditions are right, new technologies will emerge to their own rhythm, in the places and at the times most congenial to them. Leave people free to exchange ideas and back hunches, and innovation will follow. So too will scientific insight.

8

The Evolution of the Mind

Now then, listen. In order for you to fully comprehend
That minds and flimsy spirits have a birthday and an end,
I've spent long hours hunting the right words, and labour of love,
To set forth for you in poetry that's worthy of
Your life's calling. But do this favour for me just the same,
And yoke both of these concepts underneath a single name,
So that, say, when I speak of *spirit*, teaching that it dies,
Understand I am referring to the *mind* likewise,
Seeing that a single soul is formed out of their union.

Lucretius, *De Rerum Natura*, Book 3, lines 417–25

The comedian Emo Philips once joked that he considered his brain to be the most fascinating organ in his body – until he realised who was telling him this. It is a joke that brings home the absurdity of the 'self', the mind, the will, the ego or the soul. All, to the extent that they are real, are mere manifestations of the body, rather than separate from it. Yet we all talk as if a self exists, like a ghost in the machine – or, in philosopher Galen Strawson's image, as if there is a pearl of will within the shell of our body. The notion that there is a unitary piece of self-ness somewhere deep within the grey porridge inside the skull is plainly just a powerful illusion. Moreover, once you accept that the self is a bodily phenomenon, it is clear that the self is no more in charge of the body than steam is in charge of a kettle. The self is a consequence, not a cause, of thought. To think otherwise is to posit a miraculous incarnation of an immaterial spirit.

It takes a great effort to disenthrall oneself of the intentional mind, especially since it was given superficial rational legitimacy by the seventeenth-century French sage René Descartes. Descartes was not the thoroughgoing dualist he is often portrayed as. But, fairly or not, he has come to stand for the dualistic idea that there is an immaterial soul, not governed by the laws of the physical world, in our corporeal body – the pineal gland, he thought, being the place where they connected. This remained the dominant way of thinking about the mind for centuries, and persists in some forms to this day. Most of us still lazily feel as if there is a little homunculus sitting inside our head, in the front row of a sort of 'Cartesian theatre' (named after Descartes), watching the show put on by our eyes. There is a scene in the film *Men in Black* where Linda Fiorentino's character finds just such an alien homunculus, sitting at the controls inside the apparently human head of a corpse.

Yet for a while Descartes shared exile in Holland with a younger contemporary philosopher of Portuguese Jewish descent who took a far more radical, enlightened and evolutionary view. Baruch Spinoza, persecuted and exiled for his heresies, foreshadowed the conclusions of modern neuroscience to an uncanny degree. Spinoza contradicted Descartes, arguing for a devastatingly modern equality between matter and mind – for what Francis Crick later called the 'astonishing hypothesis', namely that (in Spinoza's words) 'The thinking substance [mind] and the extended substance [matter] are one and the same substance, which is now comprehended under this attribute, now under that.'

Spinoza was not, strictly speaking, a materialist, because he thought that physical events have mental causes, as well as vice versa. But he took on free will, and exposed it as at least partly an illusion. The human freedom we all boast we possess, said Spinoza, 'consists solely in the fact, that men are conscious of their own desire, but are ignorant of the causes whereby that desire has been determined'. In that sense we are no more in charge of our lives than a stone rolling down a hill is in charge of its movement.

The heretic

There is a widespread myth today that this was enough to get Spinoza anathematised as a heretic, because of the way it cast doubt upon the existence of the soul. Actually, we do not really know why Spinoza was thought heretical enough to be excommunicated from his synagogue in Amsterdam in 1656, at the age of twenty-four, because he had not yet published anything; but it is much more likely that it was for questioning the accuracy of the bible, or implying that God was part of nature. It was these heresies that caused Spinoza to be suppressed and vilified even long after his death in the same way that Lucretius had been, thus burying his scientific insights about the mind and free will.

Spinoza's *Ethics* was not published till after his death in 1677, when it caused outrage. Jews, Catholics, Calvinists and monarchs were united in disapproval. The book was banned, and copies confiscated, even in Holland. For a century the only copies were held secretly in private libraries. The only legitimate way to quote Spinoza was to surround him with derogatory epithets. When Montesquieu omitted to do this in citing Spinoza in his *l'Esprit des lois* in 1748, he was denounced and forced to recant to save his own reputation. Montesquieu's book was published in Geneva, anonymously – proof enough of the intellectual intolerance of Catholic France even long after the death of Louis XIV. When Diderot and d'Alembert's *Encyclopédie* devoted five times more space to Spinoza than to John Locke, it was muted in its praise for him, the better to disguise its heresy. Even Voltaire disparaged Spinoza with anti-Semitic jibes, showing uncharacteristic herd-following. So for a long time Spinoza never got the credit he deserves for sparking the Enlightenment.

Not only did Spinoza see the mind as a product of the emotions and urges of the body, he pointed out that even those of us motivated by impulse think we act freely:

> An infant believes that it desires milk freely; an angry child thinks he wishes freely for vengeance, a timid child thinks he wishes freely to run away. Again, a drunken man thinks, that from the free decision of his mind he speaks words, which afterwards, when sober, he would like to have left unsaid. So the delirious, the garrulous, and others of the same sort think that they act from the free decision of their mind, not that they are carried away by impulse.

It was the wine talking, says the drunkard to explain his outburst; but the sober man could just as easily say it was the lack of wine (and the influences of his parents, society and rational calculation) that made him choose not to insult his friend. In Anthony Damasio's words, 'The mind exists for the body, is engaged in telling the story of the body's multifarious events, and uses that story to optimize the life of the organism.'

Seeking homunculus

Search as you will, you cannot find the mind in the brain – or for that matter the heart – of a human body. You will find only lobes and nodules and cells and synapses – all different, all working in parallel, all talking among themselves. Whence emerges the unity of consciousness then?

At this precise moment I am thinking one thought, doing one thing, seeing one scene – but who decided that I should do that one thing from among the cacophony of other possibilities? Was there some kind of contest? I do not feel like a democratic consensus arrived at by a billion cells; I feel like a single me. And I feel as if 'I' am in charge, capable of deciding right now to think a different thought or do a different thing. I have free will – by which I mean (the definition comes from John Searle) that I could have done otherwise than I in fact did. And that, moreover, the things I could have done otherwise were neither the products of preceding forces, nor the products of random quantum swerves at the atomic level. Just as there seems to be no satisfying sense of free will in determinism, so there's no satisfying free will in randomness.

As the neuroscientist Michael Gazzaniga argues, even the most strident determinist does not actually believe he is a pawn in the brain's chess game. Yet it is indisputably the case that the conscious self is a construct, a story told after the fact to bring unity to what is actually a diverse experience. The psychologist and philosopher Nick Humphrey calls consciousness 'the magic show that you stage for yourself inside your own head'. That this is a creation or production is revealed by visual illusions, in which the brain's interpretation of what it sees goes beyond reality. Gazzaniga gives a simple demonstration of why consciousness is a post-hoc story. Touch your finger to your nose and you will experience the sensation of touch simultaneously on the nose and the finger. Yet the neural perceptions must have arrived at the brain at different times: as the neuronal impulse propagates, the fingers are three feet away, the nose just three inches away from the brain. The brain waits for both signals to arrive and integrates them into a single experience before delivering them into consciousness.

The study of the brain has found no pearl, no organ or structure that houses the self or consciousness or the will. It never will, for these phenomena are distributed among the neurons in the same way that the plan for how to make a pencil is distributed among the many contributors to a market economy. The self, says the psychologist Bruce Hood in his book *The Self Illusion*, 'emerges out of the orchestra of different brain processes like a symphony'. When people are asked to close their eyes and point to where the perception of the self originates, from the side and the front of their heads, they generally choose a point midway between the eyes and about a third of the way back from the brow ridge – not far, it has to be said, from the pineal gland that Descartes thought so vital. Open up a brain, though, and look in that spot, and you will find nothing out of the ordinary (the pineal is nothing special – a sort of hormonal way station). An alien seeking to divine the centre of the American economy would likewise probably end up in some internet-server farm in the middle of nowhere.

The astonishing hypothesis

The only conclusion, then, is that Francis Crick was right in his 'astonishing hypothesis', namely that 'A person's mental activities are entirely due to the behaviour of nerve cells, glial cells, and the atoms, ions, and molecules that make them up and influence them.' He called this idea astonishing to draw attention to how unfashionable it still was, even in the 1980s, to reject a lazy Cartesian dualism. Yet Crick's ambitious aim, as one of the two people who had stumbled upon the secret of life when he and James Watson found the self-copying code of DNA, was to find the seat of consciousness. He wanted to pin down the very structures in the brain that manifested the phenomenon of conscious, as opposed to unconscious, perception. For example, when you see an optical illusion of the kind that flips between one perception and another, such as a Necker cube, there must be some neural change as the flip happens. Where does that neural change occur?

Crick never found the answer. On his deathbed in 2004 he was correcting a paper on a structure called the claustrum, which is an especially well-connected sliver of brain tissue that's hard to experiment on, because it's so essential. But perhaps even he was still thinking in overly top–down terms. Perhaps consciousness is far too distributed among the neurons ever to be found. Earlier Crick also drew attention to the case of a patient who had suffered a lesion in Brodman's Area 24 of the brain, close to the anterior cingulate sulcus, and who had become uncommunicative, because she was unmotivated to communicate. Since a different problem, 'alien hand syndrome', in which one hand seems to take on a life of its own, was also associated with the same part of the brain, it seemed as if some sort of seat of the will had perhaps been located. Certainly, it is true that 'aboulia', or lack of motivation, is associated with damage to this part of the brain. But even if this is the location of motivation, and without it you cannot take voluntary initiative, it does not solve the philosophical conundrum. Your 'decision' to move your hand is the cause of the hand's movement, but is itself the consequence of the influences upon your brain. Area 24, in other words, is downstream of a lot of brain activity. Something gives it a nudge to initiate an action.

The most famously disturbing experiment in neuroscience was performed on people with electrodes fixed to their scalps by Benjamin Libet and colleagues twenty-five years ago. The subjects were required to push a button and register the position of an oscilloscope dot at the moment they decided to push the button. What Libet found was that although the subjects noted their decision to act two hundred milliseconds before they acted, his own electrodes picked up activity in the brain five hundred milliseconds before. In short, Libet could tell that a voluntary action was coming three hundred milliseconds before the subjects could. More recent experiments confirm the phenomenon. If you could see the activity inside a person's head when they waited to press a button on a computer keyboard, then you would know before they did what they were about to do. John Dylan Haynes and his colleagues at the Max Planck Institute in Leipzig used functional magnetic resonance imaging to measure the electrical activity of the brain and found that two regions, the frontopolar cortex and the precuneus, reliably predicted the pressing of a button ten whole seconds before the subject himself thought he took the decision.

To this a sceptic might respond that this is just because people are slow at reporting when they take a decision, but in a sense this is the very point: conscious awareness is a post-hoc report of what's going on in your head. 'You' may not be the same thing as 'conscious you'. As Sam Harris puts it: 'Am I free to change my mind? Of course not. It can only change me.'

The illusion of free will

Where does this leave free will? Many scientists, such as Gazzaniga, are quite comfortable these days calling it an illusion, albeit a powerful and even useful one. Your decision to press the button was the result of all sorts of determining forces, ranging from the experimenter's instructions to the habits you acquired as a child. What else would you like it to be a product of? Randomness? That way lies no freedom. Turning 'freedom to' into 'freedom from', Gazzaniga asks:

> What do we want to be free from? We don't want to be free from our experience of life, we need that for our decisions. We don't want to be free from our temperament because that also guides our decisions. We actually don't want to be free from causation, we use that for predictions.

The writer Sam Harris comes to the same conclusion, that free will is an illusion because 'thoughts and intentions emerge from background causes of which we are unaware and over which we exert no conscious control'. Besides, he points out, even if there were no delay between conscious and unconscious, so you cannot decide what you think till you think it, where is the freedom in that? If there is a democratic contest between impulses to decide which should be followed first, where is the freedom in that?

The biologist Anthony Cashmore has reached the same conclusions, that any action, however free it seems, 'simply reflects the genetics of the organism and the environmental history, right up to some fraction of a microsecond before any action'. What else could determine your actions but all the influences on you, external and internal? He affirms that a belief in free will is akin to religious beliefs, or to the fallacy of vitalism – the long-discredited notion that there is something physically different about the matter from which living things are made. None the less, Cashmore recognises that free will has not been assailed by scepticism among scientists as God and vitalism have been. It remains at least a convenient fiction, a skyhook from which to hang the practical necessities of the criminal justice system among other things. Perhaps, muses Cashmore, we inherit a belief in free will.

These thinkers are in the tradition of determinism that goes back at least to Spinoza. But they escape the charge so often levelled at determinists, that they are fatalists. Remember the lesson of chaos theory, that tiny differences in initial conditions can result in hugely divergent outcomes. Given that every football match starts with the same number of players, roughly the same size of pitch, the same sort of ball and the same rules, is it not astounding that every game is unique? How much more unpredictable is a human life, full of chance encounters and missed opportunities? Even two identical twins reared in the same house and educated in the same school will still be somewhat different. To be the product of all the past influences upon us is not to be destined to a particular fate in the future.

What Harris, Gazzaniga, Crick, Hood and Cashmore are influencing us to do is to abandon our prejudices and embrace the fact that we are nothing but the neural signals of our brain, multiply caused by the multiple influences upon us. Thank goodness the ego can be influenced, otherwise asking a taxi driver to take us to a hotel in an unfamiliar city would not work – the driver's behaviour and his experiences can be partly determined by you. All that determinists are asking you to accept is that there cannot be effect without cause.

However, while there is no doubt that these thinkers have banished the popular, dualist version of free will, the one that is incompatible with determinism, most philosophers refuse to concede that there is no such thing as free will. These 'compatibilists' point out that unconscious freedom originating in the body is itself a source of will, and that determinism is compatible with a form of free will. Harris argues that this is not what people mean by free will – they mean conscious will, independent of any influences upon us: where is the freedom in being outside one's own history? To Harris, compatibilism is merely an argument that some kinds of influence upon us are preferable to others: 'A puppet is free so long as he loves his strings.' In effect, or so says Harris, compatibilism is a skyhook: 'More than in any other branch of academic philosophy, the result resembles theology.' Given that one of the most prominent compatibilists is Daniel Dennett, Harris's friend and fellow horseman of the atheist apocalypse, this is, says Dennett, a low blow. Harris is in effect saying that he has found an instance in which Dennett, the scourge of skyhooks – who introduced the metaphor of the skyhook in the first place – has not gone far enough.

Not surprisingly, Dennett disagrees. While praising Harris as a brilliant clarifier of the argument against dualist free will, he says that 'Once you understand what free will really is (and must be, to sustain our sense of moral responsibility), you will see that free will can live comfortably with determinism – if determinism is what science eventually settles on.' Harris, Dennett says, is the author of his book, so why can he not be the author of his own character too? 'At what point do we get to use Harris's criticism against his own claims?' Dennett even goes so far as to accuse Harris of Cartesian dualism by shrinking 'me' to a dimensionless point, when he says that 'I, as the conscious witness of my own experience, no more initiate events in my prefrontal cortex than I cause my heart to beat.' But to say that Harris's brain does not initiate events in the prefrontal cortex is plainly false. In short, Dennett is saying that Harris has still not gone far enough in embracing free will as an emergent property of the brain.

The plain fact is that neither Dennett nor Harris is trying to make a mind-first argument, let alone rehabilitate the free will in the sense of an incorporeal spirit. Both men are making arguments for freedom as an emergent property. But what are the implications of this for responsibility?

Responsibility in a world of determinism

For many people, the chief reason for clinging to the folk version of the free-will skyhook is the same as the reason for clinging to the God skyhook or the government skyhook: to preserve the social order. Without the assumption of free will, children could not be told to work harder so they could achieve more, and murderers would seem more like victims of the influences upon them, than authors of their own actions. Anarchic excuse-making would explode, and nobody would be held accountable as society crumbled.

And up to a point this is true. The history of the Western world shows that, as we have gradually embraced bottom–up explanations, we have stopped blaming people for things that were not their fault. We once blamed ill people for the wickedness that had led them into illness; or accident victims for the sins that they were being divinely punished for. As late as the 1960s (and still today in some countries) we blamed and punished homosexuals for their inclinations, refusing to believe that they were a product of their internal influences – genetic or developmental. Today it is the fact that homosexuality is effectively innate and unwilled that is the most persuasive argument in favour of tolerance. Up till a generation ago, we blamed dyslexics for their disability, and autistics' parents for their eccentricities. We no longer do so.

We have also progressively exculpated mad people who commit violent crimes for their offences, and taken to treating them rather than punishing them. Our policy on free will has evolved away from blame.

There is no doubt that science will nudge us still farther down this road. As the neuroscientist David Eagleman has argued, the more we understand the workings of the brain at an anatomical, neurochemical, genetic or physiological level, the more we will find the causes of criminal behaviour. As we do so, we will abandon the idea of willed behaviour in many cases. The biologist Robert Sapolsky argues that our growing knowledge about the brain 'makes the notions of volition, culpability, and, ultimately, the very premise of the criminal justice system, deeply suspect'. Anthony Cashmore points out that there is no moral basis for excusing a criminal on the grounds of disease, but not excusing one on the grounds of poverty. Advances in neuroscientific knowledge will only shrink the scope of the criminal law.

But there must surely be a limit to how far we can go in this direction. Daniel Dennett argues that just because we were too punitive in the past does not make all punishment incoherent. He praises Harris's laudable motive 'to launder the ancient stain of Sin and Guilt out of our culture, and abolish the cruel and all too usual punishments that we zestfully mete out to the Guilty', punishments that are merely the human yearning for retaliation dressed up to look respectable. But Dennett then refuses to follow Harris to the logical conclusion that all punishment is unjustifiable and should be abolished: 'Punishment can be fair, punishment can be justified, and in fact, our societies could not manage without it.'

In the early 2000s, a forty-year-old Virginia schoolteacher of hitherto good character began collecting child pornography and attempting to molest his eight-year-old stepdaughter. He was sent for treatment, but his behaviour only got worse, so he was sentenced to prison. The night before the sentence started he complained of headaches and vertigo. Scans revealed a benign tumour the size of a kiwi fruit pressing on the left side of his frontal cortex. When it was removed his paedophile tendencies vanished. Some months later he began to show interest in young girls again. A part of the tumour which had been missed had regrown, and was now removed. His behaviour became normal.

In what way was this paedophile less free to act as he did than, say, a celebrity television host who molested young girls without the aid of a brain tumour? Both acted on the prompting of unconscious influences originating within their brains or else-where. Both were aware that what they were doing was wrong. We certainly see one as less 'blameworthy' than the other, but was he any less free? Sam Harris argues that once we recognise that even the most terrifying predators are unlucky to be who they are, 'the logic of hating (as opposed to fearing) them begins to unravel'.

Of course, there will be debates about different causes. Con-servatives will stress individual experience, while liberals will emphasise class circumstance. Of course, there will be abuses of our growing tendency to 'understand' rather than punish crimes – people who make spurious claims to diminished responsibility to escape harsh sentences. But does it matter greatly, so long as the public is protected? Arguably, we today incarcerate even sane murderers more for the protection of the public, and for deterrence of the crime in others, than for punishment per se. Fine.

Likewise, every time we praise somebody for overcoming a disadvantaged background to achieve great things – to go (say) from a flat above a grocery shop to become the longest-serving British prime minister of modern times despite the handicaps of gender, modest means and provincial origins – we implicitly denigrate those who do not overcome their disadvantages. Or every time we laud the courage of a cancer survivor, we imply cowardice in somebody who failed to survive. My point is that the illusion of an individual, immaterial ego, with power to take decisions, is not necessarily any more just than the opposite assumption, that each person is the sum of their influences.

So it's true that abandoning dualist free will, and embracing the idea of behaviour as an emergent property of the evolved brain, brings less judgemental attitudes, but it's far from clear that this is a bad thing. It has brought more humanity to our social policies, rather than more anarchy. Let's go the whole hog. Let's admit that free will has nothing to do with how we sentence a criminal. We treat a child who kills a parent by accident more leniently than a sadist who murders a child in a premeditated way, but not because one had more free will than the other. The murderer's action was a product of events and circumstances and genes; the child's was mainly a product of accidental events. That alters how we punish them, but it does not mean one had more free will.

Once we remove the homunculus, it becomes easier to understand freedom itself. As Dennett has argued in his book *Freedom Evolves*, 'The freedom of the bird to fly wherever it wants is definitely a kind of freedom, a distinct improvement on the freedom of the jellyfish to float wherever it floats, but a poor cousin of our human freedom.' Dennett's crucial insight is that free will is not some binary, all-or-nothing thing that you either possess or you do not. Freedom to influence your own fate is an almost infinitely variable thing that is the product of biology. The ability to move is a step towards freedom; the ability to move farther or faster is a farther or faster step. The ability to see, to hear, to smell and to think provide still more freedom to alter your fate. Technology, science, knowledge, human rights, the weather forecast – they all increase your freedom to alter your fate. It turns out that political liberty and philosophical freedom are indeed rooted in the same thing. And to appreciate them, indulge them, value them, you do not need to believe in a simplistic version of free will that is outside the material universe, any more than to celebrate the beauty of nature you need to believe it was created by a man with a long white beard, or than to benefit from the miracle of world trade you need to believe in world government. You do not need, Lucretius, to swerve.

There is a paradox inherent in seeing consciousness and free will as something that emerges and evolves from inanimate matter, namely that it makes transcendence and belief in a soul more explicable and real. As the philosopher of consciousness Nick Humphrey argues, one of the strengths of the reductionist theory is that 'it can explain how the experience of being conscious adds to people's lives by convincing them that any reductionist theory must be false'. Human beings, he thinks, are connoisseurs of consciousness who take a lot of interest in the metaphysical ramifications of being here, and that gives consciousness a genuine function. Consciousness is an 'impossible fiction' that 'works wonders to improve its subjects' lives'. The belief in the will and in the immortal soul themselves emerged as evolutionary consequences of how the brain changed. This is a far more satisfying idea than the notion that the soul and the will are real things with no history and no trace of their origin.

9

The Evolution of Personality

. . . You see,
Don't you, that even though a force outside them may propel
A crowd, sometimes stampeding them against their will, pell-mell
Yet there is something in our chest can fight back and can stand
Against it, making the mass of matter turn at its command
Throughout our body, and when that mass is spurred ahead can rein
It back into its place and settle it back down again.

Lucretius, *De Rerum Natura*, Book 2, lines 277–83

It was a tiny twist of fate – the metaphorical flap of a butterfly's wing – that set Judith Rich Harris on course to come up with an evolutionary explanation of human personality. In May 1977 she was asked by a friend, who was getting divorced, to write a classified advert for a local paper to help find a home for a dog of a rare breed. A few months later the friend, Marilyn Shaw (who was an assistant professor of psychology), remembered Harris's way with words, and asked her to help rewrite an article that had been rejected by a psychology journal. Harris, who had been ejected from the psychology PhD programme at Harvard some years before for lacking 'originality and independence', had left her job as a research assistant at Bell Labs because of ill health, so was glad to help. In editing Shaw's paper she discovered her talent for writing; and two years later, on Shaw's recommendation, she was hired by a publisher to ghost-write two chapters of an introductory psychology textbook. This led to a commission to co-author a textbook, which went through several editions; and in 1991 Harris signed a contract to write a book on her own about developmental psychology.

Only, after a bit, she stopped agreeing with what she was writing.

Psychology was then wholly in thrall to the idea that parents shaped the personalities of their children, and that differences between children were caused by parents; the only question was how. Experiment after experiment that Harris recounted demonstrated with triumphant precision the ways in which children resembled their parents for good and ill, and asserted that people were the products of what other people had done to them, especially their parents. For example, one typical study reviewed emotional expressiveness in children and found that freely expressive parents had freely expressive children, while buttoned-up parents had buttoned-up children. The authors concluded that this showed the 'socialisation of emotion'. They did not even discuss the genetic alternative: that both the parents and the children had innate tendencies to be buttoned up.

This was part of the great twentieth-century dogma of the blank slate – the notion that virtually everything in your head came from outside: not just your language, your religion and your memories, but your very character, your intelligence, your sexual preferences, your capacity to love. This dogma had conquered almost all thought during the second half of the twentieth century, not just in psychology, but in anthropology, biology, politics, and every other nook and cranny of human science. Whether you were a follower of Sigmund Freud's psychoanalysis or B.F. Skinner's behaviourism, whether you emphasised culture or diet, you were part of the same church: people were the product of others' influences. Their personalities and capabilities were inscribed on the *tabulae rasae* of their minds by influential others. This was at the time thought to be not just intellectually correct, but morally too: it meant that people were not condemned by the unfairness of their heredity. Policy was increasingly based upon the blank-slate view of human nature.

To an extent this was a reaction against the genetic determinism of the nineteenth and early twentieth centuries, when some had blamed everything on heredity, especially cultural differences between races. But the problem was that the new dogma substituted environmental determinism for genetic, and in doing so licensed just as much abuse of human rights. Communists spoke enthusiastically of moulding new forms of human nature, and looked to science for justification for vicious programmes of re-education; one (Trofim Lysenko) even insisted that he could re-educate the biology of wheat plants, and had his opponents arrested for doubting him. Moreover, the environmental determinists entrapped themselves in their own logic. Having argued that sexism and racism were wrong because there was no such thing as human nature, they made the logical inference that anybody who argued that there was such a thing as human nature must be a sexist and racist. In fact, the argument against sexism and racism, or for that matter murder, does not depend on whether sexism, racism or murder come naturally to human beings in some circumstances. These things are wrong, but not because they are unnatural.

By the 1960s the tendency to blame parental and early influence for everything had reached ridiculous extremes. Films and novels were routinely incorporating childhood traumas as singular causes of personality. Homosexuality was being blamed on hostile fathers; autism on cold mothers; dyslexia on bad teachers. Scientists who discovered flies with mutant behaviour rather than mutant anatomy were being told it was impossible, because behaviour was not in the genes. Books were being published with dogmatic titles like *Not in Our Genes*, as if DNA was entirely irrelevant. Scientists were being vilified if they argued that any part of intelligence might be heritable, or that women and men might have consistently different minds as well as bodies. If you argued that genes affected behaviour even a bit, you were pigeonholed as a heartless fatalist paving the way for a return of Nazism. By the end of the 1960s, the blank-slate dogma had conquered almost the whole of human science, and was stamping out pockets of resistance wherever they flared up in corners of academia.

But flare up they did. For a start, students of animal behaviour just could not ignore the overwhelming evidence that instinct could produce surprisingly complex behaviour. Without ever meeting its parents, a cuckoo chick knows how to eject its host's eggs from the nest, migrate to Africa, return, sing, select a victim species, and start the cycle all over again. Some of these zoologists were beginning to ask why it was that other animals should be endowed with instincts finely honed by the trial and error of natural selection on a grand scale, while human beings were instead reduced to the lottery of depending on single idiosyncratic tutors to fill their empty minds. Geneticists were beginning to notice that twins raised apart often had very similar intelligence and personality, while adopted children raised together were often very different.

When I was a student in the 1970s every such tentative suggestion of innateness in human behaviour was met with scorn and fury from the guardians of the blank-slate flame. Nature versus nurture was the flashpoint of the day, a bit like climate science today – with every heretic quickly dismissed as an extremist: How dare you say it's all in the genes! You must be some kind of Nazi sympathiser!

Powerless parents

In 1993, Judith Rich Harris was drafting her textbook on developmental psychology and obediently repeating the blank-slate nostrums of the field, when she began to have doubts about the idea that parents' actions were the source of their children's personalities through the way they doled out reward and punishment. The evidence from twin studies seemed to show that genes played a large part in determining personality; the evidence from evolutionary psychology seemed to show that universal features of human minds made evolutionary sense; and the evidence from anthropology showed that 'childrearing practices in traditional societies were nothing like what the current advice-givers were recommending, and yet the kids turned out okay'. Harris had already co-authored three editions of a textbook that hewed to the assumption that parents made personality, but she began to notice that the evidence just did not support the theory.

Children's personalities did tend to resemble those of their parents, but that could be because they shared their parents' genes. That possibility had not been ruled out in all the experiments, merely assumed away. And the differences between siblings within the same family seemed systematically incompatible with the notion that parents had put the personalities in place in each child. As Harris put it later, whenever a research method was used that controlled for genetic differences between families, then 'the home environment and the parents' style of child-rearing are found to be ineffective in shaping children's personalities'.

Harris asked to be released from the contract to write her textbook. In 1995 she published a long article in the journal *Psychological Review*, beginning with the provocative sentences: 'Do parents have any important long-term effects on the development of their child's personality? This article examines the evidence and concludes that the answer is no.' At first there was little reaction, and much of it was curious – who was this woman, with no academic affiliation or PhD? But then the American Psychological Association voted to grant Harris their George A. Miller award (worth $500) for an outstanding article in psychology. Accepting the award, Harris revealed that it was George A. Miller who had written to her to eject her from Harvard's PhD programme thirty-eight years before. Shortly after that, she set out her argument in a lengthy book, *The Nurture Assumption*, which quickly became a bestseller.

Harris pulled no punches. The importance of parenting, she said, had been overstated, and parents had been sold a bill of goods. They had a right to feel cheated. The guilt trip should stop here. Instead of helping, the nurture assumption was condemning many parents to guilt and shame when their children turned out badly. For others there was just no evidence that all the advice being doled out by educators, psychologists and agony aunts on how to raise children would make a significant difference to the child's adult personality. Being cruel or neglectful to a child was a bad thing, sure, but because it was unkind to the child, not because it left the child with a different personality. Parents mattered, sure, but because they provided care and love, not because they were the cause of differences in personality between one person and the next. Lack of parents made a difference; but different styles of parenting did not.

Meanwhile, a growing body of evidence from behaviour-genetic studies was consistently converging on the same message: that differences in personality are formed roughly half by the direct and indirect effects of genes, and roughly half by something else, which did not include the home environment at all. As Harris summarised the experiments: 'Two adopted children reared in the same home are no more similar in personality than two adopted children reared in separate homes. A pair of identical twins reared in the same home are no more alike than a pair reared in separate homes.' The child-development literature, in assuming again and again that the correlation between parents' behaviour and children's behaviour meant causation, that an abusing father makes a son into an abuser (for example), had simply failed to test the genetic explanation at all. The father's tendency to abuse might have been inherited genetically by the son. The kindness of the daughter might have been inherited from the kind mother as a disposition, not learned as a habit. And the conflict that broke up the family might not have caused the child's antisocial behaviour; far more likely that both shared the same internal cause in parent and child: the child inherited an antisocial tendency from its parents. Harris recounted a joke to make the point about confusing cause and effect in nature–nurture debates: 'Johnny comes from a broken home.' 'I am not surprised – Johnny could break any home.' Harris stresses that such 'child-to-parent' effects are common.

The reaction to Harris's book from the child-development world was as furious as you would expect when an academic discipline has its entire body of work put into question for failing to check its assumptions. A meeting arranged to discuss the book – over the strong objections of many in the field – by the National Institute of Child Health and Human Development saw Harris being openly harangued by the doyens of the discipline, especially Eleanor Maccoby and Stephen Suomi. Articles in the press castigated her for ignoring good evidence against her findings. But when she pressed for chapter and verse, the objections melted away. Suomi's strong claim that a cross-fostering experiment in calm versus anxious monkeys showed that parenting styles did influence the personalities of monkeys turned out to be untrue. He eventually admitted that the only data was from an unpublished trial with a very small number of monkeys, contradicted by other experiments that found no such effect. Jerome Kagan's claim that he had seen the same effect as Suomi in human beings (but in the opposite direction) turned out to be based on one study by a student of a small number of fearful babies followed up for just twenty-one months – hardly a lifelong effect. In short, Harris's case emerged triumphantly vindicated, rather than damaged, by everything the psychological establishment threw at her. The critics fell back on methodological criticisms of behaviour genetics, which proved largely baseless, and on much weaker claims about how parents affected children – in particular that parents treated children with different genes differently. Victory is by no means hers, and still the psychological profession and practice continues to believe in parental influence, but the idea is shrinking all the time. Children get their personalities mostly from within themselves.

The status quotient

In a follow-up book, published in 2006, Harris was able to address the really interesting mystery that the behaviour-genetics studies had revealed: what causes the 50 per cent of personality differences that cannot be caused directly and indirectly by genes? The truly bizarre thing about this difference is that it seems to be just as large in a pair of identical twins as it is in siblings or adoptees. In other words, identical twins are more similar than siblings, who are more similar than adoptees in the same family, but only because of their shared genes. Correcting for genetics, the differences that emerge between the personalities of identical twins are just as great as those between siblings and between adoptees. Whatever is causing the non-genetic differences between siblings works just as well between identical twins. Even conjoined twins are different – one is usually more outgoing and loquacious than the other, for example. Somebody whose identical twin has schizophrenia has only a 48 per cent chance of getting the disease as well.

What is the source of these huge non-genetic differences, if not parents? Harris went through five 'red herrings', and dismissed them one by one. The unexplained differences in personality could not be explained by reference to the home environment: once corrected for genetic similarities, the effect of family shrinks to zero. They could not be explained by gene–environment interactions (parents treating children differently depending on their genetic predispositions). Chance does not seem to fit the bill either. Nor different environments within the family, birth order in particular. The only large study to claim to find a consistent birth-order effect soon dissolved into an argument about unpublished data to support its assertions. Gene–environment correlation was the last red herring – bright children tending to read more books, attractive children tending to attract more attention, and so forth. This happens, of course, but it is an indirect genetic effect included in the half of personality differences attributable to genes, directly or indirectly. It's not in the bit that needs explaining.

Harris's explanation is ingenious and persuasive. She points out that human beings develop certain social systems as they mature – to socialise, to develop relationships and to achieve and recognise status. Socialisation means learning how to fit in with other people of your own age. Children acquire their habits, their accents, their favoured language, and most of their culture from their peers. They spend a lot of time learning to be similar to these peers. In forming relationships, however, they learn to discriminate between different people, adopting different behaviours with different individuals.

And then in their teens they begin to assess their relative status within their peer group. In the case of men, this mostly means working out how tall, strong and domineering you are, and adjusting your ambitions and personality accordingly. There's a fascinating finding in economics that taller men earn more money throughout their careers, but that it is their height at sixteen, not at thirty, that best predicts their earnings. The reason for this, as other studies have shown, is that this is when men decide their status, and shape their personalities accordingly. So what employers are rewarding are the attributes of self-confidence and ambition that came partly from being a tall, strong football player at school, rather than the height of the person today. Women tend to decide their status based largely on relative attractiveness, and they judge their attractiveness based on how others seem to judge them. In both sexes therefore, says Harris, there is a tendency to settle some aspects of your personality in the mid-teens, based on how high you think your relative status is amongst your peers. That, she thinks, is the likely cause of the differences in personality that are not directly or indirectly genetic.

The beauty of this explanation is that it fits identical twins well. Identical twins will differ very little in height or physical attractiveness, but they tend to differentiate their personalities from each other markedly, and outsiders pick up on this quickly and reinforce it. 'How do we tell them apart? X is the talkative one.' Even conjoined twins decide, almost arbitrarily, to be differently assertive. One is always a little more confident than the other, a difference that feeds upon itself and grows with feedback from other people. As Harris says, the status system is 'capable of producing personality differences that are unrelated to differences in genes'. You may not like Harris's emphasis on status, but that the explanation comes from within the individual, based on his or her reading of the social surroundings, is fundamentally persuasive. And your personality is yours; you are not a creature of other people. Natural selection made sure that brainwashing was not easy. And it's time we stopped looking to parenting for the credit or the blame.

The idea that parents shape their children's personalities is so ingrained, and still supplies so many psychoanalysts with their livelihoods, that any challenge to it is bound to meet a lot of resistance. Yet the evidence has been getting more and more clear: variations in personality are determined by a combination of genes and random influences, but not by parents. The central premise of Freudian analysis – that childhood events cause adult psychological problems – has been shown to stand on no good evidence whatsoever. Says Harris: 'The evidence does not support the view that talking about childhood experiences has therapeutic value.' Remember, in the early twentieth century all the advice to parents stressed discipline; in the later part of the century, all the advice stressed indulgence. Yet there is absolutely no evidence that this caused a shift in human personality in the Western world. Because people wanted there to be something they could do about our actions and tendencies, they argued that there must be an agent to blame. The nurture assumption was fuelled by many factors – worries about a return to Nazi eugenics, Rousseau-esque idealism, the doctrines of Marx, Freud and Durkheim – but the root of its appeal lay in the need to think of somebody being in charge. Instead, the truth is that personality unfolds from within, responding to the environment – so in a very literal sense of the word, it evolves.

Intelligence from within

So much for personality differences. What about intelligence? Thirty years ago it was still taboo in academia to suggest any role at all for genetics in IQ, though the person in the street had no such qualms. Today, everybody accepts the relentlessly consistent verdict of the twin studies and adoption studies: differences in intelligence owe a great deal to differences in genes. The debate is whether it is 30 per cent or 60 per cent, and whether it is mainly direct – genes creating an aptitude for learning, if you like – or indirect – genes creating an appetite for learning, and a tendency to spend time with books. As Professor Robert Plomin, probably the world expert on the genetics of intelligence, has said, there used to be a kneejerk reaction along the lines of 'You can't measure intelligence,' or 'It couldn't possibly be genetic.' Now the tone is more like: 'Of course, there is some genetic influence on intelligence, but . . .'

Many people have long feared this moment, on the grounds that it will lead to fatalism about the prospects for children, to writing off the dull ones and creating a self-fulfilling prophecy by teaching the bright ones better. Yet there is no evidence that the shift towards a more genetic view of intelligence is leading to any kind of fatalism at all. Rather, the opposite is happening, with ever more interest in coaching intelligence into the less gifted, rather than coaxing native wit from the gifted. The trend towards medicalising the things that get in the way of learning – dyslexia, attention-deficit disorder and the like – is in effect an admission that things can be innate, genetic and organic, without being irreversible.

Meanwhile, if intelligence was not significantly genetic, there would be no point in widening access to universities and trying to seek out those from modest backgrounds who have much to offer. If nurture were everything, those children who had experienced poor schools could be written off as having poor minds. Nobody thinks that. The whole idea of social mobility is to find talent in the disadvantaged, to find people who have the nature but have missed the nurture. In 2014 an article in a British newspaper criticised Boris Johnson, the Mayor of London, for believing in genetic influences on intelligence, yet its headline said that 'gifted children are failed by the system' – which assumes the existence of (genetically) gifted children.

One of the surprising things to emerge from behaviour genetics is that heritability of intelligence increases with age. The correlation between the IQs of identical twins, compared with that of adopted siblings, grows markedly as they get older. This is because families and circumstances largely determine the environments of young children, whereas older children and adults seek out and evoke the environments that suit their innate preferences, reinforcing their own natures. The longer you live, the more you express your own nature.

Still more surprising to many people is that in conditions of greater economic equality, IQ becomes more heritable, not less. In a world of more ample and more equally distributed food, obesity becomes more heritable, not less. This is because where many go hungry, fortune will largely determine who gets fat. Once everybody has enough food, the ones who get fat will be the ones with a genetic tendency to do so, and fatness will appear to run more in families – to be more heritable. The same with intelligence. Once everybody gets a similarly good education, the high achievers will increasingly be found among the children of high achievers, rather than the children of those with the best resources. It follows that a high correlation between the achievements of parents and their children, far from indicating that the parents are giving their children unfair environmental advantages, indicates instead that opportunity is being gradually levelled. Professor Plomin says that 'heritability can be viewed as an index of meritocratic social mobility', an idea that many people find counterintuitive. We are nowhere near equality of opportunity, but if we get there we will not find equality of outcome.

My point is that the new understanding that genetics matters, and that intelligence is an emergent feature of a child to be nurtured, rather than an imposition of society, is very much not to be feared. It is a meritocratic result, and presents us with a world in which people are resistant to being brainwashed because they are in charge of their own destinies. The bitter irony of the nature–nurture wars of the twentieth century was that a world where nurture was everything would be horribly more cruel than one where nature allowed people to escape their disadvantages through their own talents. How peculiarly nasty to write people off because they were born in a slum, or fostered by indifferent parents. The society depicted in Aldous Huxley's *Brave New World* is usually mistaken these days for one of fatalistic genetic determinism. In fact it is the very opposite, a place where early nurture for the elite produces unfair advantages. Fortunately, we know from the work of the economist Gregory Clark that elites regress inexorably to the mean over time. Despite sending their children to elite pre-schools, the richest of the rich in a city like New York can do little to make up for their children's genetic mediocrity; and despite getting little opportunity, brilliant kids from the slums can make it big. Nature is the friend of social mobility.

The innateness of sexuality

The confusion caused by this dawning realisation was a joy to behold. Never was the consternation of the establishment more acute than in the 1990s, when it became clear that homosexuality was much more innate and irreversible than people had been assuming, and much less a matter of early life experience or adolescent indoctrination. What a horrible conclusion! Playing into the hands of fatalists and the prejudiced, condemning people to be prisoners of their genes? Not at all. The consternation was caused by the fact that it was gay people themselves who most enthusiastically welcomed this news. See, they said, we are not perversely insisting against our natures on annoying conservatives with our homosexuality. It is what we are. It comes from within. There was a little harrumphing on the left from those who thought the new view might lead to eugenic persecution, but it soon died down when it became clear just how keen homosexuals were to be thought innately homosexual. The harrumphers on the right, meanwhile, had always justified their prejudice on the grounds that they did not want to see young people 'turned into' gays by older gays. The ground had now been pulled from under their prejudice by the realisation that you were who you were in terms of sexuality. The acceptance that homosexuality is not caused by adolescent indoctrination has played a large part in dissipating conservative opposition to gay rights.

No single episode did so much, in my view, to end the nature–nurture war as this. (I published a book on nature–nurture, a topic of ferocious debate in previous decades, in 2003; it elicited nice reviews but little interest, and the topic has been quiet ever since.) Parents could stop 'blaming' themselves or others for the sexuality of their children, and could just accept it. Gay people, and clever people and moody and cheerful people, could stop being told they were that way because of what had been done to them, and could relax in the knowledge that it was something that had emerged from inside them. It was suddenly clear that the political left should have been embracing innateness all along. The humane thing to do is to accept that human beings are built largely from within, from below; not from without, from above.

The origin of sex differences in human behaviour is a rich seam of misunderstanding about innateness and culture. Our culture relentlessly reinforces the stereotype that little boys prefer to play with trucks and little girls prefer to play with dolls. The toy shops are divided into pink girls' and blue boys' aisles, pandering to the fact that adults are quite happy to see girls and boys in conventionally different ways. This enrages many feminists, who insist that the very origin of these sex differences lies in the way they are forced upon children by the prevailing culture. But they are confusing cause and effect. Parents buy trucks for boys and dolls for girls not because they are slaves to hegemony, but because experience tells them that is what their children want. Experiment after experiment has shown that given a choice, girls will play with dolls and boys with trucks, no matter what their previous experience. Most parents are happy to reinforce sex differences, but have no interest in starting them from scratch.

In the early 2000s, the behavioural scientist Melissa Hines really put the cat among the pigeons by showing that the very same preference is true of male and female monkeys. Given the choice, female monkeys will play with dolls, males with trucks. This experiment caused fury and criticism from other psychologists determined to find fault with it. But it has since been repeated in a different species of monkey, with the same result. Female monkeys, unaware that they are slaves to cultural stereotypes, like things with faces. Male monkeys, unaware that they are doing the bidding of human sexists, like things with moving parts. In a triumphant vindication of Judith Rich Harris's argument, it has now been conclusively shown that the aisles of toy shops, with their rampant sexism, are responding to innate preferences in human beings, not causing them. These differences were not imposed, they evolved.

The evolution of homicide

If differences between people come from within, similarities do too. The ruling doctrine of the post-war period, that animals had instincts and people had learning, has also come crashing down under the realisation that evolution explains much about typical human behaviour. In virtually every mammal species, for example, the male grows larger than the female, has greater strength in its neck and front limbs, fights more often over mates or territory, is more sexually assertive, is less attentive to offspring, and shows greater variance in reproductive success (some have many children, some have none). How strange that human beings show these features too, even though people are supposedly the products of culture rather than instinct. The origin of these characteristics is not hard to discern in the fact that it is unavoidably true in mammals that, for biological reasons, females spend far more time and energy gestating and breastfeeding offspring than males spend generating sperm, so female reproductive capacity is a scarce resource for which males compete. In species where males have more surviving offspring if they stick around to help, then that habit becomes widespread – and we are one such species. Hence we show greater similarity between the sexes than do gorillas or deer. But the asymmetry between male and female habits is rarely completely lost.

One survey shows that throughout the world and throughout history, from thirteenth-century England to modern Canada, from Kenya to Mexico, men have killed men much more often – on average about ninety-seven times more often – than women have killed women. Social scientists have explained this phenomenon by reference to particular cultures: women were conditioned to be gentler; women were subordinated; women were expected to play different roles; women were more severely punished for homicide at one time (if that was once true, it is so no longer). Women and men, went the dogma, were different only to the extent that they had been treated differently by society. A leading criminologist summarised the prevailing wisdom in his profession in the 1970s when he wrote that neither biology nor psychology 'helps to explain the overwhelming involvement in crime of men over women'.

Martin Daly and Margo Wilson wrote a book on homicide in the late 1980s, and begged to differ. They argued that the cultural-determinist explanations did not fit the facts, and that it was far more likely that men were more violent for similar reasons that other male mammals were more violent – because they had in the past been forced by biology to compete for mating oppor-tunities. They pointed out that the probability of being a victim or a perpetrator of homicide is far higher in men than in women, peaks at the same age in all cultures (young adulthood), and that this is just as true in peaceable cultures with low murder rates as in violent societies with high murder rates. As a remarkable chart in the *Economist* showed in 1999, the graph of male homicides versus age, which rises rapidly in the late teens, declines steeply after peaking at twenty to twenty-five, then gradually levels off, is exactly the same shape for Chicago between 1965 and 1990 as it is for the whole of England between 1974 and 1990 – but peaks at nine hundred per million in Chicago and at thirty per million in England and Wales.

How bizarre that these facts should be universal in a species where local culture is so crucial, and how bizarre that this ten-dency to violence should peak just when males are competing most fiercely for mating opportunities, as in other mammals. The homicide statistics are dominated by young, unmarried, un-employed men seeking to improve their status or defeat sexual rivals. The same is true of hunter-gatherers and small-scale societies throughout the world: young men kill young men over women and status. Surely the explanation for most killing lies in the fact that natural selection has endowed human beings with the sort of instinct that means that (in Daly and Wilson's words) 'any creature that is recognizably on track towards complete reproductive failure must somehow expend effort, often at risk of death, to try to improve its present life trajectory'. Banish the magic of cultural determinism; look to evolution for the causes of behaviour.

The evolution of sexual attraction

Or take the startling fact that men are most attracted to women of prime reproductive age, good health and the sort of bright personality that they would most like their children to inherit. A recent study asked men and women what age they found members of the opposite sex most attractive for either short-term or long-term relationships. There was a stark difference between the sexes. Throughout their lives, women said they preferred partners of roughly their own age for both kinds of relationship. Up till the age of about thirty they preferred men who were slightly older; after that, men who were slightly younger – but even at fifty women said the most attractive age for a man was around forty-three. By contrast men of all ages (and admit it, you know what's coming!) said they found twenty-year-old women most attractive for short-term mating and for sexual fantasies. Some men in their forties nudged their preferred age up to twenty-three or twenty-four, but others stuck with twenty. For long-term mates, older men did prefer slightly older women, though still much younger than themselves. In other words, men of all ages find women of the age of maximum reproductive fertility most attractive. Seek the explanation for this not in the world of cultural norms, but in the world of evolution: men who were attracted to women of prime reproductive age and good health tended on average to leave more descendants behind than men who preferred elderly, immature, sick or morose sexual partners. Women who found strong, confident, mature and ambitious men attractive tended to leave more descendants than those who fell for weak, fearful, youthful or retiring men. It is truly strange that in my youth such explanations for universal human characteristics were *verboten*.

The Harvard psychologist Steven Pinker argues that, in stark contrast to the blank-slate dogma, our emotions and faculties have been adapted by natural selection for reasoning and communicating, have a common logic across cultures, and are difficult to erase or redesign from scratch. They come from within, not without. Learning can only happen because we have innate mechanisms to learn. Learning is not the opposite of instinct; it is itself the expression of an instinct – or rather many instincts. The human brain comes equipped automatically – though not necessarily from the start – with a tendency to learn language, to learn to recognise faces and emotions, to understand numbers, the wholeness of objects and the mindfulness of other people.

The fall of social, cultural and parental determinism and its replacement by a more balanced, evolutionary theory of human personality and character is a great liberation from an oppressive and false form of cultural creationism.

10

The Evolution of Education

Therefore we must consider well celestial happenings,
And by what principle the sun and moon run their courses,
And all phenomena upon the earth, and governing forces.
And then especially, we must nose into with sharp wits,
What makes up the soul and what the nature of it is.

Lucretius, *De Rerum Natura*, Book 1, lines 127–31

Compulsory, class-based education of young people by teachers in preparation for exams is one of those universal things nobody ever questions. We just assume that's the way learning happens. But a quick reflection on our own experience shows that there are all sorts of other ways to learn. We learn by reading, by watching, by emulating, by doing. We learn in groups of friends, we learn alone. Yet almost none of this is called 'education' – which is always a top–down activity. Is the classroom really the best way for young people to learn things? Or has the obsession with formal education crowded out all sorts of other, more emergent models of learning? What would education look like if allowed to evolve?

When you think about it, it is rather strange that liberated, free-thinking people, when their children reach the age of five, send them off to a sort of prison for the next twelve to sixteen years. There they are held, on pain of punishment, in cells called classrooms and made, on pain of further punishment, to sit at desks and follow particular routines. Of course it is not as Dickensian as it used to be, and many people emerge with brilliant minds, but school is still a highly authoritarian and indoctrinating place. In my own case, the prison analogy was all too apt. The boarding school I attended between the ages of eight and twelve had such strict rules and such regular and painful corporal punishment that we readily identified with stories of prisoners of war in Nazi Germany, even down to the point of digging tunnels, saving up food and planning routes across the countryside to railway stations. Escapes were frequent, firmly punished, and generally considered heroic.

The Prussian model

The economic historian Stephen Davies dates the modern form of the school to 1806, the year when Napoleon defeated Prussia. Stung by its humiliation, the Prussian state took the advice of its leading intellectual Wilhelm von Humboldt, and devised a programme of compulsory and rigorous education, the purpose of which was mainly to train young men to be obedient soldiers who would not run away in battle. It was these Prussian schools that introduced many of the features we now take for granted. There was teaching by year group rather than by ability, which made sense if the aim was to produce military recruits rather than rounded citizens. There was formal pedagogy, in which children sat at rows of desks in front of standing teachers, rather than, say, walking around together in the ancient Greek fashion. There was the set school day, punctuated by the ringing of bells. There was a predetermined syllabus, rather than openended learning. There was the habit of doing several subjects in one day, rather than sticking to one subject for more than a day. These features make sense, argues Davies, if you wish to mould people into suitable recruits for a conscript army to fight Napoleon.

The Prussian experiment was especially noticed across the Atlantic. Archibald Murphy, the founder of public schools in North Carolina, said in 1816 that 'The state in the warmth of her affections and solicitude for their welfare must take charge of those children and place them in schools where their minds can be enlightened and hearts can be trained to virtue.' Horace Mann, widely regarded as one of the fathers of American public education, was a keen student of the Prussian model. He visited Prussia in 1843, and came back determined to emulate that country's public schools. In 1852 Massachusetts explicitly adopted the Prussian system, followed shortly after by New York. In Mann's eyes, the purpose of public education was not mainly to raise standards (after all, by 1840 literacy in the northern states had already reached 97 per cent), but to turn unruly children into disciplined citizens. He could not have been clearer that this was for the good of the country, not the needs of the individuals. In the words of Wikipedia's entry on Mann: 'Instilling values such as obedience to authority, promptness in attendance, and organizing the time according to bell ringing helped students prepare for future employment.' It was no coincidence that America's values were at the time thought by many to be in danger of dilution by Catholic immigrants, and that this was a big part of the motive for the state taking over education. In his book *The Rebirth of Education*, Lant Pritchett quotes the frank admission of a nineteenth-century Japanese education minister: 'In administration of all schools, it must be kept in mind, what is to be done is not for the sake of the pupils, but for the sake of the country.'

Crowding out private schools

Some years later the British went down the same route, mainly to create clerks to run their empire. The British, as Sugata Mitra said in his remarkable 2013 TED lecture, set out to create a big computer with which to operate their far-flung possessions, an administrative machine made of interchangeable parts, each of which happened to be human. In order to turn out those parts, they needed another machine, an educational one, which would reliably produce people who could read quickly, write legibly, and do addition, subtraction and multiplication in their heads. As Mitra put it, 'They must be so identical that you can pick one up from New Zealand and ship him to Canada and he would be instantly functional.'

As in America, compulsory, state-mandated education was not, as many believe, the only way that learning would reach the poor. When the British state brought in compulsory education in 1880 its population was already almost entirely literate. Literacy had risen steadily from about 50 per cent among English men and 10 per cent among English women in 1700 to about 90 per cent of both sexes by 1870. In 1880, when national compulsion was enacted, over 95 per cent of fifteen-year-olds were already literate. This had come about entirely through an explosion of voluntary education within family, church and community over the preceding half-century, with the state having almost no policy on the matter before 1870. There is no reason voluntary education would not have continued to expand further in the years that followed. An entire system of education had evolved spontaneously, with no direction from government.

In 1965 Edwin West, a British economist at Newcastle University who later moved to Canada, published his now famous account of private education, *Education and the State*, in which he argued that the imposition of a state education system from 1870 in Britain, with compulsion from 1880, in effect simply displaced a growing and healthy private schooling system that would have continued to develop. In West's vivid phrase, the government merely 'jumped into the saddle of a horse that was already galloping'.

Much the same was true of India, where a survey in the 1820s found a widespread privately-funded school system reaching more boys than was the case in some European countries, long before the British introduced a public education system in the subcontinent. Mahatma Gandhi complained later that the British had 'uprooted a beautiful tree' and left India more illiterate than it had been, in displacing the indigenous private school network with a disastrously unsuccessful public one, centralised, unaccountable and open to caste exclusion. The British furiously disputed this, of course, but the evidence suggests they were wrong to do so.

Between 1818 and 1858 enrolment in private schools in England quadrupled. Education was already close to universal in Britain in 1870, though school years were short and provision was patchy by today's standards. But that's key – we cannot judge yesterday by today's standards. The system was rapidly growing, as the working class saw the advantages of being able to read the increasingly cheap and abundant printed newspapers and periodicals, and to write. As West commented: 'The belief that the appearance of the modern popular press stemmed from Forster's Education Act in 1870 is a myth . . . In the late 1860s most people were literate, most children had some schooling – and most parents were paying for it.'

It is interesting to revisit precisely what W.E. Forster recommended. Far from wanting a free and universally state-provided education system, he actually suggested that the state step in only where there seemed to be serious gaps in private provision of schooling, that fees be charged, and that parents should have choice over which school they sent their children to. All of these wishes were soon breached as the state, once given the chance, quickly took to providing almost all schooling and to deciding not just what should be taught and by whom, but which school a particular child should attend. The counterfactual, in which education continued to be a private affair after 1870, but with the state providing grants to those who could least afford the fees, would almost certainly have continued to provide an expanding and evolving education system, in which innovation and competition produced a curriculum and standards that improved just as fast as they actually did, perhaps much faster. Instead, the myth grew up that the British state stepped in where there was no education and caused the education of succeeding generations.

Such a system might well have avoided the recent deterioration of standards in state education, which has led to ever more desperate calls for affirmative action so that children from state schools can get into the best universities. Private schools disproportionately supply the best candidates for Oxford and Cambridge, a state of affairs that either indicates that the rich are innately cleverer than the poor, which seems generally unlikely, or that private schools are providing a better education, which is a shocking indictment of the quality of state education. The cost, incidentally, of an education is not much greater in the private than the public system. The difference is that the money comes from the parents in the private system, and from the taxpayers in the state system. The only cheaper option – home schooling – has an even better track record of academic achievement. The nationalisation of education provision, in short, has freed up poorer people to spend their private income (as opposed to their taxes) on other things; but it has plainly not increased their chances of social mobility – perhaps the reverse.

Innovation in education

This is true not just of Britain. An international, long-term survey of studies on 'markets versus monopolies in education' by Andrew Coulson for the Cato Institute found that both across countries and within them, 'private provision of education outshines public provision according to the overwhelming majority of econometric studies'. Lant Pritchett's devastating survey of state education in India and elsewhere found dismally low standards in many state-sponsored schools, nearly always associated with centralised control. Proud boasts that children are spending more time in schools and more money is being spent on education mean nothing if that education is failing to enable children to learn. Pritchett draws an analogy with a spider and a starfish. A spider controls everything that happens on its web through the single node of its brain: it is highly centralised. A starfish has no brain and is a radically decentralised organism with local neural control of its arms. In education, spider systems were designed in the nineteenth century, essentially to build nations to legitimate regimes. Those centralised systems are worse than useless at facing the educational challenges of today, and of innovating. Pritchett's solution is to encourage local evolution of an education system open to variety and experimentation: to make education much more like a starfish.

The real tragedy of nationalised education is how little innovation it has seen. Even when not learning Latin, and despite the fact that I was extremely well taught in some of the very best schools in the world, it was astonishing how medieval the set-up under which I was educated still was. One could not help but feel that education had not marched forward with technology in the same way that other areas of life had. Science was taught – not just to me, but to my children as well – as if it was a catalogue of facts to be regurgitated, rather than a procession of fascinating mysteries to be challenged. Give them galaxies and black holes, not Boyle's Law! It is nothing short of a miracle, said Albert Einstein, that 'the modern methods of instruction have not yet entirely strangled the holy curiosity of inquiry; for this delicate little plant, aside from stimulation, stands mainly in need of freedom'.

Nationalisation surely had a lot to do with this failure of innovation. 'It's time to admit that public education operates like a planned economy,' said Albert Shanker, long-serving President of the American Federation of Teachers. 'A bureaucratic system in which everybody's role is spelled out in advance and there are few incentives for innovation and productivity. It's no surprise our school system doesn't improve, it resembles the communist economy more than our own market economy.'

Evolutionary reform of education is happening. James Tooley, Professor of Education at Newcastle University, has catalogued – 'discovered' might be a better word – the fact that the poorest slums of cities, and the remotest villages, in countries such as India, Nigeria, Ghana, Kenya and even China abound in low-cost private schools. He first began studying this phenomenon for the World Bank in 2000 in Hyderabad in India, and has more recently followed it through Africa. In the cramped and sewage-infested slums of the old city of Hyderabad he stumbled upon an association of five hundred private schools catering to the poor. In one of them, the Peace High School, he found doorless classrooms with unglazed windows and stained walls, where children of rickshaw-pullers and day labourers paid sixty to a hundred rupees a month (about 90p–£1.50), depending on age, for their education. Yet the quality of the education was impressive. In another, St Maaz High School, he found a charismatic head teacher with mathematical flair who in twenty years had built up a school with nearly a thousand students, taught by a group of largely unqualified (but often graduate) teachers, on three rented sites, from which he made a reasonable profit. State schools existed, with state-certificated teachers in them, but many of Hyderabad's parents were exasperated by the poor quality of the education they provided, and many of the private-school teachers were exasperated by the poor quality of the teacher training. 'Government teacher training,' one told Tooley, 'is like learning to swim without ever going near a swimming pool.'

When Tooley told these stories to his colleagues at the World Bank, he was told that he had uncovered examples of business-men ripping off the poor, or that most of the private schools were creaming off the wealthier parents in a district, which was bad for those left behind. But this proved demonstrably untrue: the Peace High School in Hyderabad gave concessions, or even free tuition, to the children of extremely poor and illiterate people:

one parent was a cleaner in a mosque earning less than £10 a month. Why would such people send their children to private schools rather than to the free state schools, which provided uniforms, books and even some free food? Because, Tooley was told by parents, in the state schools teachers did not show up, or taught badly when they did. He visited some state schools and confirmed the truth of these allegations.

Tooley soon realised that the existence of these low-cost private schools in poor neighbourhoods was not unknown, but that it was largely ignored by the establishment, which continued to argue that only an expansion of state education could help the poor. The inadequate state of public education in low-income countries is well recognised; but the answer that everybody agrees on is more money, rather than a different approach. Amartya Sen, for instance, called for more government spending and dismissed private education as the preserve of the elite, while elsewhere in the same paper admitting that the poor were increasingly sending their children to private schools, 'especially in areas where public schools are in bad shape'. This bad shape, he thought, was due to the siphoning off of the vocal middle classes by private schools – rather than the fact that teachers were accountable to bureaucrats, and not to parents. Yet the poor were deserting the state sector at least as much as the middle class. The lesson that schooling can be encouraged to emerge from below was ignored in favour of the theory that it must be imposed from above.

India was just the start for Tooley. He visited country after country, always being assured that there were no low-cost private schools there, always finding the opposite. In Ghana he found a teacher who had built up a school with four branches teaching 3,400 children, charging $50 a term, with scholarships for those who could not afford it. In Somaliland he found a city with no water supply, paved roads or street lights, but two private schools for every state one. In Lagos, where government officials and the representatives of Western aid agencies all but denied the existence of low-cost private schools, he found that 75 per cent of all schoolchildren in the poor areas of Lagos state were in private schools, many not registered with the government. In all the areas he visited, both urban and rural, in India and Africa, Tooley found that low-cost private schools enrolled more students than state schools, and that people were spending 5–10 per cent of their earnings on educating their children. When he asked a British government aid agency official why his agency could not consider supporting these schools with loans instead of pouring money into the official educational bureaucracy in Ghana, he was told that money could not go to for-profit institutions.

Suppose you are the parent of a child in a Lagos slum. The teacher at the school she attends is often absent, frequently asleep during lessons, and provides a poor standard when awake. This being a public-sector school, however, withdrawing your child goes unnoticed. Your only other redress is to complain to the teacher's boss, who is a distant official in a part of the city you do not often visit; or you can wait for the next election and vote for a politician who will appoint officials who will do a better job of sending inspectors to check on the attendance and quality of teachers, and then do something about it. Good luck with that. A World Bank report cited by Tooley states despairingly that pay-for-performance cannot work in public-sector schools, and 'dysfunctional bureaucracies cascade into a morass of corruption, as upward payments from those at lower levels buy good assignments or ratings from superiors'.

If your teacher is in a private, for-profit school, however, and you withdraw your child, then the owner of the school will quickly feel the effect in his pocket, and the bad teacher will be fired. In a free system the parent, the consumer, is the boss. Tooley found that private-school proprietors constantly monitor their teachers and follow up parents' complaints. His team visited classrooms in various parts of India and Africa, and found teachers actually teaching in fewer of the government classrooms they visited than in private classrooms – sometimes little more than half as many. Despite having no public funds or aid money, the unrecognised private schools had better facilities such as toilets, electricity and blackboards. Their pupils also got better results, especially in English and mathematics.

The technology of education

The impact of profit-seeking education is not confined to poor countries. In Sweden, profit-making schools have been a competitive spur to state schools, raising standards and increasing teacher contact time. The charitable status of most elite private schools in Britain is probably getting in the way of their investment and expansion.

Technology is about to change education still more radically. The Bridge International Academies group is now running two hundred low-cost, for-profit schools in Kenya, using a syllabus scripted for the teachers and delivered by tablet computer – the computer also acting as a monitoring device to check that teachers are teaching. The idea here is that pupils should not be limited by the quality of teacher available in their district, but should get access to best practice from wherever in the world it can be supplied, via a local teacher. It's similar to the way the Khan Academy now offers more than 4,000 short videos of high-quality private tuition that anybody can use, on almost any topic. Or to the proliferation of 'massive open online courses' (MOOCs), by which top lecturers at elite universities can now be watched, and their courses taken, by thousands of eager students, not just those lucky enough to attend Stanford or MIT. Just as you do not have to listen to the local singer, but can hear Placido Domingo, so you do not have to be taught by the local teacher in the modern world. You can seek out the very best. At the other end of the spectrum is the Minerva Academy, a private college founded by the tech entrepreneur Ben Nelson in San Francisco, which is a small, indeed minimal, real university in which students live together as normal, but without all the normal features of such institutions, especially lectures, which are replaced by online, interactive seminars. Lectures, says Minerva's Stephen Kosslyn, are 'a great way to teach, but a terrible way to learn'.

The traditional university will surely be gone in fifty years, swept away by technology. Why pay huge fees to spend three years on one campus, earning the right to be paid not very much more in the real world than non-graduates, rather than putting together your own combination of online courses, marked and graded online, using the lectures of the best teachers in the field wherever they happen to be? When Sebastian Thrun, an artificial-intelligence expert, sent out one email announcing that he would teach a course not just to his students at Stanford but to whoever wanted to listen in on the internet, tens of thousands took the course. Over four hundred of them got better grades than the top student at Stanford.

In fact, why not cut out the human almost entirely? When Sugata Mitra first put a computer with online access in a hole in a wall in a Delhi slum, he did not know what to expect. He watched as children crowded round the screen and began to play with the internet. Within weeks, he found, kids who did not even speak English had bootstrapped their way into surprisingly deep expertise.

The hole-in-the-wall experiment sparked the idea behind the film *Slumdog Millionaire*. In three years, Mitra's colleagues found, twenty computers made 6,000 children in one part of New Delhi computer-literate without any teaching. Children could learn to use computers without adult instruction. Crucially, they were not learning by themselves, but teaching each other: it was a collective, emergent phenomenon.

In Mitra's mind this discovery soon sparked the thought that other kinds of learning could happen without teaching in a connected world. He set up an experiment in a school in a remote Tamil-speaking village called Kalikuppam, near Pondicherry:

to teach ten-to-fourteen-year-old children who knew almost no English and even less biology the rudiments of molecular biology, but to do so without any instructor who knew biology. In just two months the children taught themselves biotechnology to an average test score of 30 per cent. The procedure was to give them access to a hole-in-the-wall computer and present them with well-posed questions, then leave them to their own devices.

This experiment, now repeated worldwide, led to the concept of the 'self-organised learning environment', or SOLE. Mitra insists that three, four or five children share each computer, then poses them questions and leaves them to find out the answers. Who was Pythagoras? How does an iPad know where it is? What was the British Raj? Can trees think? Why do we dream? Were Vikings smelly? Each question, says Mitra, triggers arguments, but invariably opens the door to learning.

Curiously, in some ways Mitra may be rediscovering a method that is ancient and Indian, but has lain buried beneath layers of Prussian practice. In the late 1700s a British teacher named Andrew Bell, working in Madras, discovered that Indian schools used older boys to teach younger boys, with remarkable success. He brought the idea back to Britain, introduced it into numerous schools and published an influential book entitled *An Experiment in Education, made at the Male Asylum at Madras; suggesting a System by which a School or Family may teach itself, under the Superintendence of the Master or Parent.*

Mitra's next step was to invent the 'granny cloud', a network of usually retired people in Britain who give online guidance to children in schools in remote villages or slums. 'I had a new hypothesis,' wrote Mitra. 'Given the appropriate digital infrastructure, a safe and free environment, a friendly but unknowledgeable mediator, groups of children can pass their school-leaving tests . . . on their own.'

One of the biggest obstacles to self-organised learning, Mitra thinks, is the assessment system. So long as exams remain tests of memory and mental regurgitation, there is little point in self-education, and schools will be prevented from evolving new forms. A recent exam question in Britain, for instance, was 'What is an oxbow lake?' Think about this for a moment. In the days when district commissioners were paddled by natives down rivers to administer justice, this might have been a useful fact to know in advance. Today, for the vanishingly few people who might need to know what an oxbow lake is, the answer is available at the click of a smartphone. If the exams had questions like 'What is self-similarity, and what are the latest findings in this area?', Mitra told me, then there would be no choice but to allow the use of the internet in the examination hall, and that would change everything.

Indoctrination continues

We need to get away from creationist thinking in education, and allow it to evolve. Education, done properly, is an emergent, evolutionary phenomenon. It is the process of encouraging learning about the world. Yet it is also a tool of propaganda and indoctrination, of what John Stuart Mill called 'despotism over the mind'. Even when they had stopped thinking of their products as cannon fodder, or as barbarians in need of civilising, nationalised schools did much to teach children well into the twentieth century that their country was glorious and usually right, while its rivals were perfidious and usually wrong, that God was a Christian, and so forth. True, there is less of that particular propaganda in the curriculum today, though some policy-makers are troubled about what goes on in schools dominated by radical Islamists. But there is, if anything, more of another kind of propaganda. It may be the gospel of multiculturalism and respect for the planet, so it is 'good' indoctrination, but it's still indoctrination. You do not need to be a wild-eyed conspiracist to see far less opening of minds than training people in what to think going on in modern schools. Platitudes about the state of the world, or the desirability of wind energy, seem to crop up with alarming frequency in children's textbooks, even when the ostensible topic is history or Spanish.

A recent report, written by Andrew Montford and John Shade, found that the British educational curriculum in 2014 was so intent on teaching children to be environmental activists that it was replete with 'serious errors, misleading claims, and bias through inadequate treatment of climate issues in school teaching materials. These include many widely-used textbooks, teaching-support resources, and pupil projects.' Textbooks and teaching materials suggest writing to politicians, joining campaigns, and badgering parents. The phrase 'global warming' turns up in exam papers on economics, chemistry, geography, religious studies, physics, French, humanities, biology, citizenship, English and science.

Fortunately, children do not always do what adults tell them. And indoctrination is not new. Doris Lessing once wrote that we should say to children: 'You are in the process of being indoctrinated. We have not yet evolved a system of education that is not indoctrination. We are sorry, but it's the best we can do.' There is one education system that seems to be better than most at inculcating resistance to indoctrination, at least in the early years. Montessori schools, with their collaborative, test-free, mixed-age classrooms and emphasis on self-directed learning, have a remarkable track record in producing entrepreneurs. The founders of Amazon, Wikipedia and Google (both of them) went to Montessori schools. The secret, according to Larry Page of Google, may lie in the schools' habit of bringing out children's natural tendency for 'not following rules and orders, and being self-motivated, questioning what's going on in the world, doing things a little bit differently'.

Education to deliver economic growth

The very purpose of education has been distorted all too often by a top–down fantasy. Rarely, if ever, has the purpose of state education been to add to scholarship and generate knowledge. The purpose instead is to train an obedient citizenry, loyal to the nation, likely to deliver economic growth and brainwashed with the latest fashion in ideology. 'The aim of public education is not to spread enlightenment at all. It is simply to reduce as many individuals as possible to the same safe level, to breed a standard citizenry to put down dissent and originality,' said H.L. Mencken. That's partly why the dismal lack of innovation and progress in education has never mattered much to those in power. Today, in Stephen Davies's view, schools are little more than devices for signalling to employers that a young person has been sufficiently indoctrinated to stick to a task and do as he is told, just as Horace Mann would have wanted. Left-wing politicians tend to emphasise the spending of money, while right-wing ones emphasise reform of the curriculum and methods of teaching. But both agree that education is a national, not an individual priority. Any effect it might have on the individual is secondary to its impact on the country. Ask not what your country's schools can do for you . . .

In the past twenty-five years, the main obsession of governments, aside from inculcating anxiety about the state of the planet into the next generation, has been to use education to deliver economic competitiveness. It has been an assumption right across the political spectrum that better schools, better universities, better vocational education and better training will deliver a more prosperous society. It is certainly true that longer-educated individuals are more prosperous – more education leads to higher salaries. And it is also true that highly educated countries are generally more prosperous. But do the facts bear out this notion that education is the elixir of economic growth? Is there any evidence that it was education that drove countries to prosperity, or vice versa? Alison Wolf examined the data in exhaustive detail in her book *Does Education Matter?*, and concluded that the answer is a surprising 'no'. She points to World Bank studies that show a negative relationship between education levels and growth. The countries that devoted the most resources to expanding their education systems grew less rapidly than those that devoted fewer resources to education. Egypt did a great job of improving, lengthening and spreading education, but grew slowly. In the thirty years from 1970 the country more than doubled both its secondary and its university enrolment. Yet over that period it rose only from the forty-seventh to the forty-eighth poorest country in the world. The Philippines had a much higher literacy rate than Taiwan in 1960, but today has one-tenth the income per head. Argentina, one of the least successful economies of the past century, had one of the highest literacy rates. The more a country adopted central planning, the better its education system did, but the worse its economic performance – not least because, like Egypt, it churned out many would-be bureaucrats trained to do the central planning.

Vocational education, by the way, ought to be better but rarely is. You would think it would be dominated by the needs of the sectors of the economy that are its clients. But amazingly, a different report into vocational education by Alison Wolf found that it is centralised and *dirigiste*: 'Vocational education has been micro-managed from the centre for decades. This is a bad idea, and not just because it is inherently ineffective. It also means that government takes direct public responsibility for success and failure, and finds it correspondingly impossible to be honest.'

It is true that well-educated people tend to be richer than poorly-educated people, both within and between nations. But as Wolf argues, cause and effect are confused here. 'Could it be,' she asks, 'that growth causes education rather than education causing growth?' You can certainly find examples of countries that deliberately planned and achieved great improvements in education, both basic and vocational, and which grew very fast: South Korea is the classic case, Singapore another. But Wolf asks if the education caused the growth, or was even a critical factor, and concludes that it may not have been. Hong Kong and Switzerland grew just as fast, but with far less central planning or investment in education. Switzerland has far lower than average enrolment in universities for an economy of its level. Hong Kong's 'meteoric economic growth had nothing to do with centrally planned education policy', concludes Wolf. Instead, Hong Kong parents began pushing their children into good private schools once they got rich enough to do so.

A bigger example lies across the Pacific from Hong Kong. For decades America has consistently performed poorly in international league tables of educational achievement in school, yet performed well in economic terms. The countries with the most education simply do not show greater productivity growth than the ones with less. Each year spent in school or university should be enabling an employee to be more productive, but there is no sign of this in the economic statistics. As Wolf concludes: 'If high-quality schooling is making any difference to the relative economic performance of countries, it is doing so in a very undramatic fashion, since its effects appear to be swamped or neutralized by other factors.' Education clearly benefits the individual's earning power, but it does not determine the growth rate of the whole economy.

Far from seeing an economic dividend from education, Wolf finds that countries that boosted their education levels the most tended to grow more slowly than countries that neglected to increase their education spending so much. Her conclusion is stark: 'The simple one-way relationship that so entrances our politicians and commentators – education spending in, economic growth out – simply doesn't exist.' She concedes of course that some education is necessary. Without good literacy and numeracy, it would not be possible for most well-paid jobs to exist. That is not the issue. Rather it is whether, beyond a certain level, more education – let alone more education spending – does more good. 'The idea that having the most education will get you the most prosperity is a chimera,' says Wolf. A great many jobs today are open only to graduates, although evidence suggests that they could be perfectly well handled by non-graduates.

Remember, this is very much not saying that higher education is not a good thing for the individual. It is a wonderful thing, but it is one of the rewards of economic growth, not one of the drivers. And obviously a total lack of education would be catastrophic for a modern economy; but that's not the same as saying that the best way to improve the economy is to spend more on education. Education is not a skyhook from which to hang economic policy; it is an emergent phenomenon.

Education is dominated by creationist thinking. The curriculum is too prescriptive and slow to change, teachers are encouraged to teach to the exam rather than to the pupils' or their own strengths, the textbooks are infused with instructions about what to think instead of how to think, teaching methods are more about instructing than learning, the possibilities of self-organised learning are neglected, government domination of schooling is accepted without question, and education spending is justified in terms of what it supposedly does for the country rather than the individual. None of this is meant to imply that education would happen without schooling, that teachers need not exist, that child-centred learning in primary schools is the answer, or that some kind of government policy on education is not desirable. Of course these things matter. But there is a path not taken, in which politicians and teachers both allow best practice to evolve and emerge, in which the state acts as enabler rather than dictator, in which students are encouraged to learn rather than be told what to think, in which the eager learner is boss, not servant, of the system.

Let education evolve.

11

The Evolution of Population

Sooner or later, you will seek to break away from me,
Won over by doomsayer-prophets. They can, certainly,
Conjure up for you enough of nightmares to capsize
Life's order, and churn all your fortunes with anxieties.

Lucretius, *De Rerum Natura*, Book 1, lines 102–5

For more than two hundred years, on the topic of human population, a disturbingly vicious thread has run through Western
history, basing itself on biology and justifying cruelty on an
almost unimaginable scale. When I began researching this book
I thought of Malthusian theory, eugenics, Nazi genocide and
modern population control as separate and distinct episodes in
human history. I am no longer so sure. I think there is some
persuasive evidence that a direct, if meandering, intellectual
thread links the Poor Laws, the Irish famine, the gas chambers
of Auschwitz and the one-child policies of Beijing. In all cases,
cruelty as policy, based on faulty logic, sprang from a belief that
those in power knew best what was good for the vulnerable and
weak. Urgent ends justified horrible means. Evolution was taken
as a prescription for interference, not a description of an emergent process.

Parson Robert Malthus (he's often called Thomas these days, but in his lifetime he used his middle name, Robert) casts a long shadow over the past two centuries. A wealthy English mathematician, teacher and clergyman with a fine literary style, he is known today for just one short document, the *Essay on Population*, first published in 1798 and frequently revised in the following years. He is a bit of a hero to many in the environmental movement to this day for his emphatic insistence that there are limits to growth – that population growth must lead to misery, starvation and disease when the land, the food, the fuel or the water runs out. According to his epitaph in Bath Abbey, he was noted for 'his sweetness of temper, urbanity of manners and tenderness of heart, his benevolence and his piety'. He was clearly not a nasty man, and his chief remedy for overpopulation – late marriage – was not a cruel one. But he did, none the less, think that cruel policies would, if late marriage could not be taught, be effective at halting population growth: we would have to encourage famine and 'reprobate specific remedies for ravaging diseases'.

Unfortunately, this cruel lesson is what most people have taken from Malthus: that you have to use unkind means to justify kind ends. This trope – that being kind to poor and ill people is a bad idea – runs right through eugenic and population movements, and is alive and well even now. When I write or speak about falling child mortality in Africa today, I can be sure I will get a response along exactly these Malthusian lines: but surely it's a bad thing if you stop poor people dying? What's the good of bringing economic growth to Africa: they will only have more babies – and more cars. Better to be cruel to be kind. Let's call it Malthusian misanthropy. And it is 180 degrees wrong. The way to get population growth to slow, it turns out, is to keep babies alive, to bring health, prosperity and education to all.

There were plenty, in his lifetime and afterwards, who thought Malthus's recommendations cruel. Friedrich Engels called Malthusianism a 'vile and infamous doctrine'. Pierre-Joseph Proudhon called it 'the theory of political murder; of murder from motives of philanthropy and for love of God'.

The Irish application of the theory

Yet Malthusian doctrines influenced policy directly and frequently during the nineteenth century – usually without the emphasis on the age of marriage. Britain's new Poor Law of 1834, which attempted to ensure that the very poor were not helped except in workhouses, and that conditions in workhouses were not better than the worst in the outside world, was based explicitly on Malthusian ideas – that too much charity only encouraged breeding, especially illegitimacy, or 'bastardy'. The Irish potato famine of the 1840s was made infinitely worse by Malthusian prejudice shared by the British politicians in positions of power. The Prime Minister, Lord John Russell, was motivated by 'a Malthusian fear about the long-term effect of relief', according to a biographer. The Lord-Lieutenant of Ireland, Lord Clarendon, thought 'doling out food merely to keep people alive would do nobody any good' (not even the recipient of the food?). The Assistant Secretary to the Treasury Charles Trevelyan had been a pupil of Malthus at the East India Company College: famine, he thought, was an 'effective mechanism for reducing surplus population' and a 'direct stroke of an all-wise and all-merciful Providence' sent to teach the 'selfish, perverse and turbulent' Irish a lesson. Notice the Malthusian misanthropy, and the invocation of the ultimate skyhook, Providence. Trevelyan added: 'Supreme Wisdom has educed permanent good out of transient evil.' We are back to Dr Pangloss and the Lisbon earthquake: mass death is a good thing. In short, a million Irish starved to death as the result of a deliberate act of Malthusian policy, at least as much as through ecological misfortune.

For those, like me, brought up to think of British imperialism as generally benign compared with other forms of the habit, the story gets worse. As recounted by Robert Zubrin in his book

Merchants of Despair, in 1877 a dreamy and Bohemian poet with an opium habit named Robert Bulwer-Lytton was serving as Viceroy of India, sent there by his friend the Prime Minister, Benjamin Disraeli. Bulwer-Lytton may sound like a harmless if high-born hippie, but unfortunately he was a Malthusian – or his advisers were. A drought afflicted some parts of the country. There was still plenty of food in India as a whole – food exports doubled and doubled again in two years – but taxes and the devaluation of the rupee left the hungry unable to afford relief. Bulwer-Lytton quoted almost directly from Malthus in his response: 'The Indian population has a tendency to increase more rapidly than the food it raises from the soil.' His policy was to herd the hungry into camps where they were fed on – literally – starvation rations (slightly fewer calories per head than would be handed out in Nazi concentration camps), with the result that 94 per cent died each month. Bulwer-Lytton specifically halted several private attempts to bring relief to the starving. The justification of his government was that it was being cruel to be kind, because Malthusian ends justified harsh means. Up to ten million people died.

Malthus's impact on history was not all bad. He heavily influenced Charles Darwin and Alfred Wallace. But even Darwin, that most gentle and compassionate of men, was at least briefly tempted by the idea that his beloved natural selection should be a prescription rather than a description. In an explicitly Malthusian passage in the *Descent of Man* he notes that the 'imbecile, the maimed and the sick' are saved by asylums and doctors; and that the weak are kept alive by vaccination. 'Thus the weak members of civilized species propagate their kind,' something that any cattle breeder knows is 'injurious to the race'. He goes on to lament the fact that 'the very poor and reckless, who are often degraded by vice, almost invariably marry early, whilst the careful and frugal, who are generally otherwise virtuous, marry late in life'. It was not much of a policy suggestion, and it was a rare lapse in a carefully apolitical career, but the passage clearly echoes the prescriptive Malthusian doctrines that Darwin absorbed as a young man.

Nationalising marriage

It was a hint that was enthusiastically embraced by several of Darwin's followers, notably his cousin Francis Galton and his German translator, Ernst Haeckel. Galton wanted people to choose their marriage partners more carefully, so that the fit would breed and the unfit would not. 'What nature does blindly, slowly, and ruthlessly,' he argued, 'man may do providently, quickly and kindly.' He also wanted the 'childish' 'negro' displaced from his native continent of Africa by the slightly less stupid 'Chinaman', and he thought Jews were 'specialised for a parasitical existence upon other nations'. Even for his time Galton could be a censorious and prejudiced man, though he never actually recommended sterilisation or killing of 'unfit' people.

Galton's followers were soon outdoing each other in their prescriptive rush to nationalise marriage, license reproduction and sterilise the unfit. Many of the most enthusiastic eugenicists, such as Sidney and Beatrice Webb, George Bernard Shaw, Havelock Ellis and H.G. Wells, were socialists, who thought the power of the state would be necessary to implement this programme of selective human breeding. But plenty of politicians right across the political spectrum, from Winston Churchill to Theodore Roosevelt, became keen advocates of eugenic intervention in the private lives of their citizens. Indeed, it became politically incorrect in elite circles in Britain, France and the United States not to urge eugenic policies. To be against eugenics was to be uncaring about the future of the human race.

In Germany, Haeckel took the Malthusian struggle in a quasi-religious direction, trying to fuse Darwinian with Christian insights in a theory that he called Monism. In a lecture at Altenburg in 1892 he used phrases from both Malthus and Thomas

Hobbes: 'Here it was Darwin, especially, who thirty-three years ago opened our eyes by his doctrine of the *struggle for existence*, and his theory of selection founded upon it. We now know that the whole of organic nature on our planet exists only by a relentless *war of all against all*' (my emphasis). Haeckel's followers gave eugenics a racial tinge. They advocated not just legalised infanticide for abnormal children, and systematic murder for the purpose of racial improvement, but warfare as 'the highest and most majestic form of the struggle for existence' (the words are those of Otto Ammon, writing in 1900). There's that exact phrase, *struggle for existence*, first used by Malthus in Chapter 3 of his population essay, and then by Darwin to describe the lesson he took from Malthus ('I happened to read for amusement Malthus on Population, and being well prepared to appreciate the struggle for existence which everywhere goes on . . .'). It's a phrase that, thanks to the Monists, would soon be used to justify both the Kaiser's and Hitler's belligerence. German militarists of the period before the First World War made disturbingly frequent references to Darwin, but so did those in other countries. An article in the Royal United Services Institution journal in 1898 asked: 'Is not war the grand scheme of nature by which degenerate, weak or otherwise harmful states are eliminated . . . ?' The Italian futurist Filippo Marinetti called war 'the sole hygiene of the world'.

In 1905 four of Haeckel's followers founded the German Society for Racial Hygiene, a step that would lead pretty well directly to the Nuremberg laws, the Wannsee conference and the gas chambers. It is therefore not at all hard to trace a clear path that leads from Malthus's followers' insistence that we intervene in selective survival to the ash of Birkenau. This is not to blame the innocent mathematician-clergyman for the sins of the Nazis. There is nothing morally wrong in describing a struggle for existence as a feature of human population. What is wrong is prescribing it as deliberate policy. The sin that is committed at every step is one of active interference, of ends justifying means. As

Jonah Goldberg wrote in his book *Liberal Fascism*: 'Almost all the leading progressive intellectuals interpreted Darwinian theory as a writ to "interfere" with human natural selection. Even progressives with no ostensible ties to eugenics worked closely with champions of the cause. There was simply no significant stigma against racist eugenics in progressive circles.'

It mattered little that scientific support for this policy was weak in the extreme. In fact the discoveries of Gregor Mendel, which became known to the world in 1900, ought to have killed eugenics stone dead. Particulate inheritance and recessive genes made the idea of preventing the deterioration of the human race by selective breeding greatly more difficult and impractical. How were those in charge of breeding the human race supposed to spot the heterozygotes who carried but did not express some essence of imbecility or unfitness? How long were we supposed to go on weeding out the unfit as they emerged from marriages of heterozygotes? It would take centuries, and along the way the problem would become worse as our species became more and more inbred, allowing more and more homozygous combinations. Yet the genetic facts made no difference to the debate. Driven by a planning fantasy, the political classes, left and right, agitated to nationalise reproduction to prevent the spread of unfit bloodlines.

The First International Congress of Eugenics assembled in London in 1912 under the presidency of Leonard Darwin, son of Charles. It was attended by three ambassadors, as well as the Lord Chief Justice and the First Lord of the Admiralty – one Winston Churchill. In his presidential address Leonard Darwin made no bones about the switch from description to prescription: 'As an agency making for progress, conscious selection must replace the blind forces of natural selection.' Fortunately, Britain, the birthplace of the eugenic movement, never enacted a specifically eugenic law, thanks largely to a bloody-minded Member of Parliament, Josiah Wedgwood, who spotted the danger and filibustered a eugenic Bill in the House of Commons.

Sterilisation begins

In the United States it was a different story. The Eugenics Record Office, established at Cold Spring Harbor, New York, in 1910 by the energetic eugenicist Charles Davenport, with funding from the widow of the railroad magnate E.H. Harriman, soon began to exert a powerful influence on policy. The Second International Congress of Eugenics assembled in New York in 1921, under the honorary presidency of Alexander Graham Bell, presided over by the President of the American Museum of Natural History, Henry Fairfield Osborn, and with the invitations sent out by the State Department. This was no fringe event. Leonard Darwin was too unwell to attend, but sent a message expressing the 'firm conviction . . . that if wide-spread eugenic reforms are not adopted during the next hundred years or so, our Western Civilization is inevitably destined to such a slow and gradual decay as that which has been experienced in the past by every great ancient civilization'.

The Eugenic Records Office director Harry Laughlin drew up a model eugenics law in 1932. This, together with his and Davenport's energetic lobbying, eventually persuaded thirty states to pass laws allowing for the compulsory sterilisation of the feeble-minded, insane, criminalistic, epileptic, inebriate, diseased, blind, deaf, deformed and dependent. By the time such laws were struck down in the early 1970s, some 63,000 people had been forcibly sterilised and many more persuaded to accept voluntary sterility.

It was not long before another strand of thinking entered the stream of eugenic misanthropy: nature worship. A 1916 book by a New York lawyer and conservationist, Madison Grant (founder of the Bronx Zoo, the Save-the-Redwoods League and Denali National Park), entitled *The Passing of the Great Race*, hymned the virile virtues of Nordic types and the threat to their dominance from Mediterranean and East European immigration. The book influenced the passage of the 1924

Immigration Act. It also became the 'bible' of Adolf Hitler, or so he wrote enthusiastically to Grant.

In Germany too the preservation of nature went hand in hand with the destruction of human life. 'Ask the trees, they will teach you how to become National Socialists!' read one Nazi slogan. Nazis often railed against modern farming methods, idealised closeness to nature and sang the praises of organic, peasant agriculture. Their favourite philosophers, such as Martin Heidegger, waxed lyrical about living in harmony with nature: 'Saving the earth does not master the earth and does not subjugate it, which is merely one step from boundless spoliation.' As Martin Durkin has observed, green thinking was no mere sideline for the Nazis:

> It was the green attempt of the Nazis to recreate a peasant society which led them to invade Poland in search of 'living space'. It was their green nostalgia for the Middle Ages which led to their 'blood and soil' racist ideology. It was their green anti-capitalism and loathing of bankers which led them to hate Jewish people.

In 1939 the American social reformer Margaret Sanger set up the 'Negro Project', intended to bring birth control to black people with the help of ministers and doctors. The project was unabashed in its eugenic racism: 'The mass of significant Negroes still breed carelessly and disastrously, with the result that the increase among Negroes [is] in that portion of the population least intelligent and fit.'

California was especially enthusiastic about eugenics. By 1933 it had forcibly sterilised more people than all other states combined. So when the Third International Congress of Eugenics gathered at the American Museum of Natural History in New York in 1932 under the presidency of Charles Davenport, and Davenport asked, 'Can we by eugenical studies point the way to produce the superman and the superstate?', it was to California that the superman-worshipping German delegates looked for an answer. One of them, Ernst Rudin of the German Society of Racial Hygiene, was elected to head the International Federation of Eugenics Organisations. Within months, Rudin would be appointed Reichskommissar for eugenics by the incoming Nazi government. By 1934, Germany was sterilising more than 5,000 people per month. The California conservationist Charles Goethe, who like Madison Grant combined a pioneering passion for protecting wild landscapes with an equal passion for sterilising psychiatric patients without their consent, returned from a visit to Germany overjoyed that the Californian example had 'jolted into action a great government of 60 million people'. Germany took its racism from its own Haeckelian tradition, but it got the practical knowhow of sterilisation from the west coast of America.

Justifying murder

What happened next has not lost its power to shock. Nazi Germany sterilised 400,000 people in the six years after Hitler came to power, including schizophrenics, depressives, epileptics, and disabled people of all kinds. It forbade sexual intercourse between Jews and non-Jews, and then began systematically to persecute Jews in multiple ways. Under pressure of propaganda, many ordinary Germans inverted their consciences, becoming ashamed of any feelings of sympathy with their Jewish friends: the morally correct thing to do, they thought, was to override such feelings – Malthusian misanthropy again.

Jewish immigration from Germany was actively resisted by the governments of Britain, France and America, often with explicit eugenic justification. A Bill to allow 20,000 Jewish children above the quota into the United States was defeated in Congress in early 1939 by a coalition of nativist and eugenic pressure groups assembled by Harry Laughlin. In May 1939 the ship *St Louis*, carrying 937 German Jews, sailed for Cuba.

Laughlin demanded that America not lower its 'eugenical and racial standards', and as the ship awaited permission to dock in Havana, it was announced that the passengers' US visas had been invalidated. Most of them were returned to Europe, where many were killed.

In 1939 the Nazi government established a programme called Aktion T4, which went a step further and began to kill disabled and mentally ill people, mainly with lethal injections. The first to be killed were children with congenital illnesses, 5,000 of whom were put to death. Then 70,000 adults were killed under the programme, before protests from relatives supposedly put a halt to it in 1941. Far from being the end of it, this simply led to a new plan – to herd the 'unfit' into concentration camps for mass extermination, along with homosexuals, Gypsies, political prisoners and millions of Jews. Six million human beings died. To argue that this would not have happened if Malthus, Darwin, Haeckel and Laughlin had not lived goes too far. Yet the explicit justification for the Nazi genocide was based on eugenic science stemming from the struggle for existence outlined originally by Malthus.

Population again

After the Second World War, with the revelation of the horrendous results of these policies taken to extremes, eugenics fell from fashion. Or did it? Surprisingly quickly and surprisingly blatantly, the very same arguments resurfaced in the movement to control world population. The son of the prominent pre-war eugenicist Henry Fairfield Osborn, also named Henry Fairfield Osborn, published a book in 1948 entitled *Our Plundered Planet*, which revived Malthusian concerns about the rapid growth of the human population, the depletion of resources, the exhaustion of soil, the overuse of DDT, an excessive reliance on technology and a rush to consumerism. 'The profit motive, if carried to the extreme,' wrote the wealthy Osborn, 'has one certain result – the ultimate death of the land.' Osborn's book was reprinted eight times in the year it was published, and translated into thirteen languages.

At almost the same time William Vogt, a biologist driven by a passion for wildlife conservation, published a very similar book, *Road to Survival*, in which the ideas of the 'clear-sighted clergyman' Malthus were even more explicitly endorsed. 'Unfortunately,' wrote Vogt (yes, unfortunately!), 'in spite of the war, the German massacres, and localised malnutrition, the population of Europe, excluding Russia, increased by 11,000,000 people between 1936 and 1946.' In India, he thought, British rule had contributed to making famines ineffectual, which was a pity because it led to more babies, or to Indians 'breeding with the irresponsibility of codfish'.

Fairfield Osborn established and chaired the Conservation Foundation, which tapped into his contacts to build up a large funding programme that supported many of today's big environmental groups, including the Sierra Club, the Environmental Defense Fund, and in Europe the World Wildlife Fund. His cousin Frederick Osborn had served as Treasurer of the Third International Eugenics Conference, and continued to serve as President of the American Eugenics Society. The Planned Parenthood Foundation was founded in 1916 by Margaret Sanger, who thought philanthropy would 'perpetuate constantly increasing numbers of defectives, delinquents, and dependents'. The organisation's international arm was headquartered in the offices of the British Eugenics Society as late as 1952. The population-control movement was, to an uncomfortable extent, the child of the eugenics movement.

The link was just as explicit on the other side of the Atlantic. In 1952 Sir Charles Galton Darwin, nephew of Leonard and grandson of Charles, who was a distinguished physicist, published his own pessimistic book, entitled *The Next Million Years*. 'To summarize the Malthusian doctrine, there can never be more people than there is food for,' he wrote. 'Those who are most anxious about the Malthusian threat argue that the decrease of population through prosperity is the solution of the population problem. They are unconscious of the degeneration of the race implied by this condition, or perhaps they are willing to accept it as the lesser of two evils.' Darwin was arguing that population growth could never be brought under control except by drastic means – by warfare, infanticide, or the sterilisation of a fraction of the adult population, which he feared would be 'vehemently resisted'. He simply could not foresee a happy ending to the population explosion, because he was thinking in top–down terms. How do 'we' solve the problem?

Sir Julian Huxley, the first head of UNESCO and an early advocate of population control, played a pioneering role in the environmental movement in Britain, somewhat like Osborn in America. His pre-war enthusiasm for eugenics remained un-dimmed as late as 1962, when he said at a Ciba Foundation meeting on the subject of 'Man and his Future':

> At the moment the population certainly wouldn't tolerate compulsory eugenic or sterilization measures, but if you start some experiments, including some voluntary ones, and see that they work and if you make a massive attempt at edu-cating people and making them understand what is at issue, you might be able, within a generation, to have an effect on the general population.

Sir Charles Galton Darwin, Sir Julian Huxley, Henry Fairfield Osborn Jr and William Vogt were not outliers, ignored by an embarrassed intelligentsia. They caught the mood of the times and were immensely influential.

Population blackmail

By the 1960s these ideas had converted many people in pos-itions of power. Osborn and Vogt's books had been read by a generation of students, including Paul Ehrlich and Al Gore. The most influential disciple was General William Draper, whose commission on foreign aid reported to President Eisenhower in 1959 that aid should be tied explicitly to birth control, in order to decrease the supply of recruits to communism. Eisenhower did not buy this, and nor did his Catholic successor John F. Kennedy.

But Draper did not give up. His Population Crisis Committee gradually won over many of the most influential people in American public life to the thesis that coercive population control was essential to defeating communism. Eventually, with the help of a RAND Corporation study which argued (using an absurd 15 per cent discount rate) that children had negative economic value, Draper and his allies won Lyndon Johnson's endorsement in 1966, and population control became an official part of American foreign aid. Under its ruthless director Reimert Ravenholt, the Office of Population grew its budget till it was larger than that of the rest of the entire US aid budget. In a series of alarming episodes, Ravenholt bought up defective birth-control pills, unsterilised intra-uterine devices and unapproved contraceptives for distribution as aid in poor countries. He made no bones about his views that the prevention of infant mortality in Africa was *enormously harmful to African societies when the deaths prevented thereby are not balanced by prevention of a roughly equal number of births* . . . Many infants and children rescued from preventable disease deaths by interventionist programs during the 1970s and 1980s have become machete-wielding killers' (emphasis in original).

With Ravenholt in charge of the Office of Population, and World Bank President Robert McNamara refusing loans to countries that did not submit to sterilisation quotas set for them by the bank, it became inevitable that countries like India had to undertake coercive sterilisation just to receive food aid. When in 1966 Indira Gandhi arrived in Washington to beg for food aid to relieve the famine in India caused partly by the recent war with Pakistan, she was told by Secretary of State Dean Rusk that 'a massive effort to control population was a condition for receiving the aid'. She got the message, and agreed to state by state quotas for sterilisation and IUDs. Hundreds of sterilisation camps were set up, in which paramedics performed vasectomies, IUD insertions and tubectomies by the thousand. The pitiful rewards paid to those who submitted to these procedures – twelve to twenty-five rupees per sterilisation – were enough to attract millions of hungry people, especially among the poorest. The number of sterilisations reached three million a year by 1972–73.

Some Western commentators thought starvation was a better course of action. William and Paul Paddock wrote a bestseller in 1967 called *Famine 1975!*, which argued that a time of famine was imminent, and food aid was futile. America, they said, must divide the underdeveloped nations into three categories: those that could be helped, the walking wounded which would stagger through without help, and 'those so hopelessly headed for or in the grip of famine (whether because of overpopulation, agricultural insufficiency, or political ineptness) that our aid will be a waste; these "can't-be-saved nations" will be ignored and left to their fate'. India, Egypt and Haiti should be left to die in this way.

A year later, Paul Ehrlich's *The Population Bomb* was almost as callous. India could never feed itself, Ehrlich decided. An unabashed advocate of coercion to achieve population control, he compared humanity to a cancer, and recommended surgery: 'The operation will demand many apparently brutal and heartless decisions. The pain may be intense.' Population control at home would require 'compulsion if voluntary methods fail'. He suggested adding sterilants to the water supply to achieve 'the desired population size'. As for overseas, he wanted food aid to India made conditional on forcible sterilisation of all those who had three or more children: 'coercion in a good cause', he called it. The fact that President Johnson's linking of aid to India with population control had sparked criticism at home left Ehrlich 'astounded'. In a book written jointly with his wife Anne and John Holdren, now science adviser to President Obama, Ehrlich recommended that a 'planetary regime' be 'given responsibility for determining the optimum population for the world and for each region for arbitrating various countries' shares within their regional limits'.

When Mrs Gandhi asked the World Bank for loans in 1975, she was told that stronger efforts to control India's population were needed. She turned to coercion, her son Sanjay running a programme that made many permits, licences, rations and even housing applications conditional on sterilisation. Slums were bulldozed and poor people rounded up for sterilisation. Violence broke out repeatedly. In 1976, when eight million Indians were sterilised, Robert McNamara visited the country and congratulated it: 'At long last India is moving effectively to address its population problem.'

The population sceptics

Yet here is the astounding thing. Birth rates were already falling in India and elsewhere. Food production was rising far faster than population, in a reverse of Malthusian predictions – thanks to synthetic nitrogen fertiliser and new short-strawed varieties of cereals: the Green Revolution. The answer to the population explosion turned out not to be coercion, or the encouragement of infant mortality, but the very opposite. By far the best way to slow down population growth was to keep babies alive, because then people would have fewer of them as they planned smaller families.

And the even more shocking fact is that this evolutionary solution was already known to some at the very start of the panic. Even at the very birth of neo-Malthusian population alarm in the 1940s there were some who saw how horribly wrong both the diagnosis and the cure were. Far from more babies causing more hunger, they argued that it was the other way round. People increased their birth rate in response to high child death rates. Make them richer and healthier and they would have fewer babies, as had already happened in Europe, where prosperity had led birth rates down, not up. As Earl Parker Hanson wrote in his book-length response to William Vogt, *New Worlds Emerging*, the solution to both food shortage and too many babies was prosperity, not Malthusian starvation. People would be 'more likely to think about having fewer children when they are in a position to worry about sending them to college'.

The Brazilian diplomat Josué de Castro, in his book *The Geopolitics of Hunger*, was even bolder in his criticism of the neo-Malthusians, saying that 'The road to survival, therefore, does not lie in the neo-Malthusian prescriptions to eliminate surplus people, nor in birth control, but in the effort to make everybody on the face of the earth productive.'

In the 1970s Paul Ehrlich's brand of population pessimism was attacked by the economist Julian Simon in a series of articles and books. Simon argued that there was something badly wrong with a thesis that the birth of a baby is a bad thing, but the birth of a calf is a good thing. Why were people seen as mouths to feed, rather than hands to help? Was not the truth of the past two centuries that human wellbeing had improved as population had expanded?

Famously, in 1980 Simon challenged Ehrlich to a bet about future prices of raw materials. Ehrlich and a colleague, eager to take up the offer, chose copper, chromium, nickel, tin and tungsten as examples of materials that would grow scarcer and more expensive over the next ten years. Simon bet against him. Ten years later, grudgingly and while calling Simon an 'imbecile' in public, Ehrlich sent Simon a cheque for $576.07: all five metals had fallen in price in both real and nominal terms. (One of my proudest possessions is the Julian Simon Award, made from those five metals.) Simon offered another bet to anybody who would take it: 'I'll bet a week's or a month's pay that just about any trend pertaining to material human welfare will improve rather than get worse.' Nobody took up his offer before Simon's untimely death in 1998.

The solution to the population explosion turned out to be the Green Revolution and the demographic transition. Emergent phenomena rather than coercion and planning. Evolution, not prescription. It was an evolutionary, spontaneous and unplanned phenomenon that slowed population growth. Unexpected, unpredicted and unheralded, people started having smaller families because they were richer, healthier, more urban, more liberated and more educated. Not because they were told to. There is only one country where population control was sufficiently coercive to achieve its end – China – yet all that China achieved was a deceleration of population growth almost exactly the same scale as other countries that used almost no coercion at all.

The Western origins of the one-child policy

China's one-child policy surely has little to do with the Western Malthusian tradition. Not so. That policy derives directly from the neo-Malthusian writing, and disturbingly it is probably the first and most far-reaching policy ever to be initiated by scientists. The precedent is not encouraging for those of us who love science.

Despite the fact that he presided over much suffering by the Chinese people, Mao Zedong's approach to population was relatively restrained and humane: known as 'Later, Longer, Fewer', it encouraged lower fertility by delaying marriage, spacing births and stopping at two, but in a flexible and non-prescriptive way. That is roughly what Malthus himself had advocated. Whether for that reason or because of decreasing child mortality, China's birth rate halved between 1971 and 1978. Then after Mao's death came a turn to a much more rigid and prescriptive approach. As Susan Greenhalgh, a Harvard anthropologist, recounts in her book *Just One Child*, in 1978 Song Jian, a guided-missile designer with expertise in control systems, attended a technical conference in Helsinki. While there he heard about two books by neo-Malthusian alarmists linked with a shadowy organisation called the Club of Rome. One was *The Limits to Growth*, the other *A Blueprint for Survival*.

Founded in the 1960s by an Italian industrialist and a Scottish chemist, the Club of Rome is a talking shop for the great and the good, devoted to the worship of Malthus, and meeting behind closed doors at lavish venues. Together with its affiliates it still attracts leading names, from Al Gore and Bill Clinton to the Dalai Lama and Bianca Jagger. 'The real enemy, then, is humanity itself,' the Club of Rome declaimed in a book in 1993, and 'democracy is not a panacea. It cannot organize everything and is unaware of its own limits.' In 1974 in its second report, called 'Mankind at the Turning Point', the Club of Rome issued a call for creationist thinking that remains unparalleled in its technocratic arrogance:

> In Nature organic growth proceeds according to a Master Plan, a Blueprint. Such a 'master plan' is missing from the process of growth and development of the world system. Now is the time to draw up a master plan for sustainable growth and world development based on global allocation of all resources and a new global economic system.

The Limits to Growth sold ten million copies, and purported to prove with computer models that humanity was doomed because of overpopulation and the exhaustion of resources. The book forecast that several metals could run out by 1992, helping to precipitate a collapse of civilisation and population in the subsequent century.

Written by wealthy British businessman Sir Edward Goldsmith but signed by a veritable Who's Who of the scientific establishment including Sir Julian Huxley, Sir Peter Medawar and Sir Peter Scott, *A Blueprint for Survival* rather gives the lie to the idea that the environmental movement was a grassroots, radical thing. It was motivated by the elite's habitual dislike of change, technology and consumerism. The book positively oozes snobbish disdain for the fact that consumer society, with its 'shoddy' goods, is coming within reach of ordinary people. In saying this was a mistake, it was telling the rich what they wanted to hear. Few of 'us', according to *A Blueprint for Survival*, take into account the 'dull and tedious work' that has to be performed in order to manufacture domestic appliances that supposedly save time for women in the home. As for the global poor, 'it is unrealistic to suppose that there will be increases in agricultural production adequate to meet forecast demands for food'. The authors then command that governments must acknowledge the population problem, 'and declare their commitment to ending population growth; this commitment should also include an end to immigration'. It is a highly reactionary document, of the kind that would embarrass a fringe right-wing party today.

These were the two books that that Song Jian, the father of the one-child policy, picked up in Helsinki. *The Limits to Growth* had applied control systems theory, of the kind Song was an expert in, not to the trajectory of missiles but to the trajectory of population and resource use. Song returned to China, where he republished the main themes of both books in Chinese under his own name, and shot to fame within the regime. He quickly recognised, thanks to his military experience, that (in the words of the anthropologist Susan Greenhalgh) 'The one-child-for-all policy both assumed and required the use of big-push, top–down approaches in the social domain.' Song was proposing social engineering in the most literal sense. Vice-Premier Wang Zhen was an immediate convert on reading Song's report, and put it in front of Chen Yun and Hu Yaobang, senior lieutenants of Deng Xiaoping himself. Deng apparently liked the fact that Song argued that Chinese poverty was caused by overpopulation, not economic mismanagement, and was bamboozled by the mathematics into not questioning his assumptions. At a conference in Chengdu in December 1979 Song silenced critics who were worried about the humanitarian consequences, and persuaded the party to accept his calculation that China needed to reduce its population by about one-third by 2080 in order to live within its ecological means.

General Qian XingZhong was put in charge of the policy. He ordered the sterilisation of all women with two or more children, the insertion of IUDs into all women with one child (removal of the device being a crime), the banning of births to women younger than twenty-three, and the mandatory abortion of all unauthorised pregnancies right up to the eighth month of pregnancy. Those who tried to flee and have babies in secret were tracked down and imprisoned. In some cases their communities were fined, encouraging betrayal of neighbours. The brutal campaign of mass sterilisation, forced abortion and infanticide was exacerbated by the voluntary murder of baby girls on a genocidal scale as parents tried to ensure that their one legal child was a boy. Fertility fell, but not much faster than it would have done if a policy of economic development, public health and education had been adopted instead.

What was the international reaction to this holocaust? The United Nations Secretary General awarded a prize to General Qian in 1983, and recorded his 'deep appreciation' for the way in which the Chinese government had 'marshalled the resources necessary to implement population policies on a massive scale'. Eight years later, even though the horrors of the policy were becoming ever more clear to all, the head of the United Nations Family Planning Agency said that 'China has every reason to feel proud of its remarkable achievements' in population control, before offering to help China teach other countries how to do it. A benign view of this authoritarian atrocity continues to this day. The media tycoon Ted Turner told a newspaper reporter in 2010 that other countries should follow China's lead in instituting a one-child policy to reduce global population over time.

Malthus's poor laws were wrong; British attitudes to famine in India and Ireland were wrong; eugenics was wrong; the

Holocaust was wrong; India's sterilisation programme was wrong; China's one-child policy was wrong. These were sins of commission, not omission. Malthusian misanthropy – the notion that you should harden your heart, approve of famine and disease, feel ashamed of pity and compassion, for the good of the race – was wrong pragmatically as well as morally. The right thing to do about poor, hungry and fecund people always was, and still is, to give them hope, opportunity, freedom, education, food and medicine, including of course contraception, for not only will that make them happier, it will enable them to have smaller families. Abandon the creationism of technocratic pessimism, the repeatedly debunked doom-mongering of the scientific elite with its simplistic and static misunderstanding of the nature of resources, the easy resort to the lazy plural pronoun 'we' and the dreadful word 'must'. Embrace instead the evolutionary, unplanned, emergent phenomenon of the demographic transition.

Leave the last word to Jacob Bronowski, speaking at the end of his television series *The Ascent of Man*. Standing in a pond at Auschwitz-Birkenau, where many of his relatives died, he reached down and lifted up some mud: 'Into this pond were flushed the ashes of some four million people. And that was not done by gas. It was done by arrogance, it was done by dogma, it was done by ignorance. When people believe that they have absolute knowledge, with no test in reality, this is how they behave. This is what men do when they aspire to the knowledge of gods.'

12

The Evolution of Leadership

It is far preferable to live in peace and to obey
Than to wish to reign in power and hold whole kingdoms in your sway.
Let others wear themselves out all for nothing, sweating blood,
Battling their way along ambition's narrow road.

Lucretius, *De Rerum Natura*, Book 5, lines 1129–32

In Denis Diderot and Jean d'Alembert's *Encyclopédie*, the manifesto of the French Enlightenment, you will find almost no entries for named people. To read a short biography of Isaac Newton, for example, you must look up 'Wolstrope', an old name for the Lincolnshire village where Newton grew up. There was a reason for this strange coyness. Diderot and his friends thought that history had given too much credit to leading people, and too little to events and circumstances. They wanted to take kings, saints and even discoverers down a peg. They wanted to remind their readers that history is a process driven by thousands of ordinary mortals, not ordained by a few superhuman heroes. They wanted to remove the skyhooks from history as well as from government, society and science. (However, not even they could find anything to say about Wolstrope other than that it was Newton's birthplace.)

Diderot's older contemporary, Charles, Baron de Montesquieu, also insisted that leaders were taking credit for inexorable inevitabilities of nature. Human beings, he thought, were mere epiphenomena: history was driven by more general causes. 'Martin Luther is credited with the Reformation,' he wrote. 'But it had to happen. If it had not been Luther it would have been someone else.' The chance result of a battle could bring forward or delay the ruin of a nation, but if the nation was due to be ruined it would happen anyway. Montesquieu thus made the distinction between ultimate and proximate causes that became such a useful concept in social science. At times he became an excessive climatic determinist, as he sought inanimate causes for events, but it is little wonder he so annoyed the Church and state, who preferred God and king to get the credit.

In the nineteenth century, under the influence of Thomas Carlyle's 'Great Man' theory of history, biography was back. Carlyle said that heroes like Napoleon, Luther, Rousseau, Shakespeare and Mohamed were cause, not effect, of the times they lived in. The influential 1911 edition of the *Encyclopaedia Britannica* goes to the opposite extreme from the *Encyclopédie*: social history is buried within biography. So to read about the post-Roman world you must look up the entry on Attila the Hun.

Largely in vain had the philosopher Herbert Spencer* fought

---

* Herbert Spencer is one of the most unfairly traduced figures of history, whose reputation today is that of a heartless social Darwinist happy to let the devil take the hindmost. This is a calumny of the first order. He advocated sympathy, compassion and charity for those left behind in life's race, and he championed competition because it raised the standard of living for all, not because it helped the most successful. He was a subtle and brilliant thinker of great compassion and liberalism. An ardent opponent of militarism, imperialism, state religion, state tyranny and all forms of coercion, he was a feminist and a proponent of organised labour. So it is diametrically wrong to accuse him of thinking that 'might is right'. But it is true that he deplored the path of his contemporary rival, Karl Marx, towards seeing the state as the means of liberation. He always distrusted government, which he feared would 'play the tyrant when it should have been the protector', and preferred to encourage voluntary cooperation.

back against top–down history, arguing that Carlyle was wrong. Leo Tolstoy too devoted a part of *War and Peace* to an argument against the Great Man theory. But then the twentieth century seemed to prove Carlyle right, as great men and women – for good and ill – changed history repeatedly: Lenin, Hitler, Mao, Churchill, Mandela, Thatcher. As Boris Johnson, the Mayor of London, has argued in his book *The Churchill Factor: How One Man Made History*, it is almost impossible to conceive of any other British politician close to power in May 1940 who would have chosen not to negotiate with Hitler in search of peace, however humiliating. Nobody else in the War Cabinet had the courage, the insanity, the sheer effrontery to defy the inevitable and fight on. As Johnson makes the case, this really is a sure example of one person changing history. So is history driven by great men?

The emergent nature of China's reform

I am not so sure. Consider the reform of China's economy that began under Deng Xiaoping in 1978, leading to an economic

---

His cynical view of the state was surely amply borne out by the events of the twentieth century and the hundred million corpses left by communism. As Deirdre McCloskey put it, 'Anyone who after the twentieth century still thinks that thoroughgoing socialism, nationalism, imperialism, mobilization, central planning, regulation, zoning, price controls, tax policy, labor unions, business cartels, government spending, intrusive policing, adventurism in foreign policy, faith in entangling religion and politics, or most of the other thoroughgoing nineteenth-century proposals for governmental action are still neat, harmless ideas for improving our lives is not paying attention.' Nor did Spencer advocate heartlessness towards the unfortunate. So it is unfair that his reputation today comes largely from a hostile and misleading account written by a Marxist historian, Richard Hofstadter, in 1944, at the height of Western, as well as Eastern, enthusiasm for authoritarian policy. See Richards, Peter 2008. Herbert Spencer (1820–1903): Social Darwinist or Libertarian Prophet?, *Libertarian Heritage* 26 and Mingardi, Alberto 2011. Herbert Spencer. Bloomsbury Academic. Deirdre McCloskey's remarks are from an essay called 'Factual Free-Market Fairness'. Available at bleedingheartlibertarians.com.

flowering that raised half a billion people out of poverty. Plainly, Deng had a great impact on history and was in that sense a 'Great Man'. But if you examine closely what happened in China in 1978, it was a more evolutionary story than is usually assumed. It all began in the countryside, with the 'privatisation' of collective farms to allow individual ownership of land and of harvests. But this change was not ordered from above by a reforming government. It emerged from below. In the village of Xiaogang, a group of eighteen farmworkers who despaired at their dismal production under the collective system and their need to beg for food from other villages, gathered together secretly one evening to discuss what they could do. Even to hold the meeting was a serious crime, let alone to breathe the scandalous ideas they came up with.

The first, brave man to speak was Yen Jingchang, who suggested that each family should own what it grew, and that they should divide the collective's land among the families. On a precious scrap of paper he wrote down a contract that they all signed. He rolled it up and concealed it inside a bamboo tube in the rafters of the house. The families went to work on the land, starting before the official's whistle blew each morning and ending long after the day's work was supposed to finish. Incentivised by the knowledge that they could profit from their work, in the first year they grew more food than the land had produced in the previous five years combined.

The local party chief soon grew suspicious of all this work and this bountiful harvest, and sent for Yen, who faced imprisonment or worse. But during the interrogation the regional party chief intervened to save Yen, and recommended that the Xiaogang experiment be copied elsewhere. This was the proposal that eventually reached Deng Xiaoping's desk. He chose not to stand in the way, that was all. But it was not until 1982 that the party officially recognised that family farms could be allowed – by which time they were everywhere. Farming was rapidly transformed by the incentives of private ownership; industry soon followed. A less pragmatically Marxist version of Deng might have delayed the reform, but surely one day it would have come. And the point was that it came from an ordinary person, as Diderot would have expected. 'The moral of this story is that autocrats get too much credit for episodes of increased economic freedom,' wrote William Easterly.

One cannot, of course, say the same about Mao Zedong. The overwhelming and enormous harm he did to the Chinese people over several decades did indeed begin at the top. The collectivisation of agriculture, the extraction of grain from starving peasants to pay for nuclear weapons, the crazy plan to smelt metals in villages during the Great Leap Forward, the vicious vendettas against individuals during the Cultural Revolution – these were indeed the actions of a 'Great Man' in all the wrong senses of the phrase. As Lord Acton said, great men are mostly bad men.

Mosquitoes that win wars

Today we are still in thrall to Great Man history, if only because we like reading biography. American presidential politics is entirely based on the myth that a perfect, omniscient, virtuous and incorruptible saviour will emerge from the New Hampshire primary every four years, and proceed to lead his people to the promised land. Never was this messianic mood more extreme than on the day Barack Obama won the presidency. This was the moment, he himself had said in June 2008, when 'the rise of the oceans began to slow and our planet began to heal'. He was going to 'heal this nation', close Guantánamo Bay, reform healthcare, bring peace to the Middle East. He was given the Nobel Peace Prize simply for having been elected. Amid such expectations he could not, poor chap, fail to disappoint. As Andrew Bacevich, a political scientist at Boston University, commented in 2013, amid the disappointments of Obamacare's launch, 'Obama himself may have turned out to be something of a dud, but the cult of presidential personality that has dominated American politics for decades now still persists.' Doomed every four years to disappointment when a demigod turns out to have feet of clay, when the most powerful man in the world turns out not to have much power to change the world, the American people none the less never lose faith in the presidential religion. It is not much different in other countries.

Or pull back and observe that the great sea changes in human history – the Renaissance, the Reformation, the Industrial Revolution – occurred as accidental by-products of other things. Trade made Italian merchants rich, and because they felt guilty about committing usury, they commissioned artists to produce pious works of incomparable beauty, and supported open-minded inquiries into the learning of the classical world. Printing made possible the cheap and widespread dissemination of texts, which enabled religious reformers, after several abortive attempts in previous centuries, to undermine the authority of the Pope and his henchmen. As the technology expert Steven Johnson has argued, the unintended consequences of historical events can be far-reaching. Gutenberg made printed books affordable, which kicked off an increase in literacy, which created a market for spectacles, which led to work on lenses that in turn resulted in the invention of microscopes and telescopes, which unleashed the discovery that the earth went round the sun.

In *1493*, his magnificent account of the great Columbian exchange that followed contact between the eastern and western hemispheres, Charles Mann shows how again and again the forces that truly shaped history came from below, not above. For instance, the American Revolution was won by the malaria parasite, which devastated General Charles Cornwallis's army in the Carolinas and on the Chesapeake Bay, at least as much as it was won by George Washington. I say this not as a bad-loser Brit seeking excuses, but on the authority of the distinguished (American) environmental historian J.R. McNeill. Referring to female mosquitoes of the species *Anopheles quadrimaculatus*, he writes: 'Those tiny amazons conducted covert biological warfare against the British army.'

In 1779 the British commander Henry Clinton adopted a 'southern strategy', and sent his forces by sea to occupy the Carolinas. But the Carolinas were infested with malaria, which broke out afresh every spring, especially among new arrivals from Europe. It was the *vivax* strain of the parasite, which debilitates and weakens its victims, sometimes facilitating death from other causes. Rice-growing made the problem worse, providing ample habitat for the mosquitoes. 'Carolina in the spring is a paradise, in the summer a hell and in the autumn a hospital,' wrote one German visitor. Most white colonists had survived malaria in their youth and acquired some resistance. Most black slaves had brought some degree of genetic immunity to malaria with them from Africa. So the American south was the worst possible place to invade with foreign troops.

After capturing Charleston, the British under Cornwallis marched inland. As platoons of perspiring, pale-skinned Scotsmen and Germans tramped through the woods and rice fields in June 1780 (the height of the mosquito season), the *Anopheles* mosquitoes and *Plasmodium* parasites could not believe their luck. They both gorged themselves on blood, the mosquitoes swallowing it, the parasites being swallowed by its cells. When the time came to fight a battle, most of the army was debilitated by fever, including Cornwallis himself. In the words of McNeill, Cornwallis's army simply melted away at one battle. Only the local loyalists, fever-seasoned, could stay in the field. It did not help that the only cure for malaria – quinine from the bark of the Cinchona tree – was monopolised by the Spanish, who had cut off trade with the British in support of their French and American allies.

Come the winter, Cornwallis's men recovered, and he moved them north to inland Virginia, away from the coastal swamps, specifically to 'preserve the troops from the fatal sickness, which so nearly ruined the Army last autumn'. But Clinton ordered him to return to the coast ready to receive reinforcements, so Cornwallis reluctantly fell back to Yorktown, a fort between two feverish swamps on the Chesapeake Bay. George Washington, with French and northern troops, marched south to besiege him, arriving in September. Cornwallis, his 'force daily diminished by sickness', surrendered within three weeks. Since malaria takes more than a month to incubate, the newly arrived French and Americans began to fall sick only after the battle was over. 'Mosquitoes,' says McNeill, 'helped the Americans snatch victory from the jaws of stalemate and win the Revolutionary War, without which there would be no United States of America. Remember that when they bite you next Fourth of July.'

Of course, we cannot take away all the credit that George Washington earned as a general. But the American leaders' reputations were made by the turn of events at least as much as the other way round. Microscopic events, at that. Of course, you could argue that the war was unwinnable for the British anyway, and that eventually they would have succumbed even without the mosquitoes. It is important not to substitute a Great Insect theory for a Great Man theory. But then, that rather reinforces the point that the determinants of war were bottom–up ones.

Imperial chief executives

The Great Man theory lives on as strongly as ever in one field of human endeavour: big business. Even in the age of the internet, most modern companies are set up like feudal fiefs, with a king in charge; or a god invested with a near supernatural reputation, a very large shareholding and a reverberantly hard name like Gates, Jobs, Bezos, Schmidt, Zuckerberg. Surely it is the height of irony that the most iconic, powerful and imperial chief executives are found today in companies that float in the fluid, egalitarian, dynamic world of the digital economy. Their firms provide cobwebs of horizontal interaction among billions of customers, their employees wear jeans, eat vegan salads and work flexible hours. Yet the pronouncements of their bosses are treated as scripture. Jeff Bezos's favourite saying is 'Start with the customer and work backwards,' but it is repeated as a mantra so frequently by his staff that you cannot help thinking they start with the boss and work forwards. At the death of Steve Jobs in 2011 it was widely assumed that the survival of Apple itself was at risk, and the share price plunged. Did even Genghis Khan have this sort of effect when he died? Why has the autocratic ethos of Henry Ford and Attila the Hun survived unchanged into the twenty-first century in this way? Why are companies still such top–down things?

California technology companies originally set out to be self-consciously different in this respect from the snobbish and hier-archical businesses of the east coast and the old world. As Tom Wolfe documented as long ago as the 1980s, people like Robert Noyce of Intel deliberately intended to escape from the feudal model of east coast capitalism, with its 'vassals, soldiers, yeomen and serfs, with layers of protocol and perquisites, such as the car and driver, to symbolize superiority and establish boundary lines'. Noyce did not even have a reserved parking space at Intel. Symbols of democratic flatness do persist in west coast com-panies, and chief executives behave less like feudal overlords – but more like oracles, prophets or deities, their pronouncements treated with reverence.

As the economist Tom Hazlett put it to me after reciting some of the wide-eyed optimism being expressed about the new sharing economy we are supposedly inventing: 'There sure are a lot of billionaires in the new wiki-economy.' In filing for Face-book's initial public offering in 2012, Mark Zuckerberg stated his desire that the world's information infrastructure should be 'a network built from the bottom up, or peer-to-peer, rather than the monolithic, top–down structure that has existed to date'. Steven Johnson points out that Zuckerberg none the less con-trols 57 per cent of the company's shares, and comments wryly that 'top–down control is a habit that will be hard to shake'.

Shake it we must. As Gary Hamel wrote in an article in the *Harvard Business Review* in 2011, paraphrasing a butcher in Shakespeare's *Henry VI*, 'First, let's fire all the managers.' He points out that layers of management increase in number, size and complexity as organisations grow larger, because managers need managing too; and that a large part of a boss's job in a big firm is to keep an organisation from collapsing under the weight of its own complexity. Prescriptive management means a far greater risk of foolish decisions: 'Give someone monarch-like authority, and sooner or later there will be a royal screwup.' It also means slower decisions as problems get bounced between dilatory committees. And it disempowers junior staff, who think nobody listens to their concerns or suggestions. As Hamel points out, a person who is free to make a $20,000 purchase of a car as a customer, might not be free to buy an office chair for $500 as an employee. Little wonder that big companies grow more slowly than small ones (firms whose chief executives attend the annual World Economic Forum schmooze-fest in Davos tend to underperform the stock market), and big public bodies have worse reputations than small ones.

For all his or her apparent power, the chief executive of a big firm these days is sometimes little more than a hired spokesman. He is perpetually on the road, explaining 'his' strategy to investors and customers, relying on a chief of staff or two to hire, fire, promote or ostracise his people. Sure, there are some who do actually inculcate their philosophy deep into the organisation and design its products themselves. But they are the exceptions. Most CEOs are along for the ride, paid well to surf on the waves their employees create, taking occasional key decisions, but no more in charge than the designers, middle managers and above all customers who choose the strategy. Their careers increasingly reflect this: brought in from outside, handsomely rewarded for working long hours, then ejected with little ceremony but much cash when things turn sour. The illusion that they are feudal kings is maintained by the media as much as anything. But it is an illusion.

So who does run a company these days? Not the shareholders or the board. They largely find out after the fact that things have gone well or badly. Nor are firms cooperatives. Anybody who has tried to run a company by consensus will tell you how disastrously bad an idea that is. Interminable meetings follow hard upon each other's heels as everybody tries to get everybody else to see his or her point of view. Nothing gets done, and tempers fray. The problem with consensus is that people are not allowed to be different. It's like trying to drive a car in which the brake and the accelerator have to do similar jobs. No, what really works inside a big firm is division of labour: you do what you're good at, I'll do what I'm good at, and we'll coordinate our actions. That is what actually happens in practice inside most companies, and good management means good coordination. The employees specialise and exchange, just like participants in a market, or citizens in a city.

The evolution of management

A Californian firm called Morning Star Tomatoes has been experimenting with 'self-management' for two decades. The result is that Morning Star is the largest processor of tomatoes in the world, handling 40 per cent of California's processed tomato crop. Its profits have grown rapidly, it has very low employee turnover, and is highly innovative. Yet it has no manager, no bosses and no chief executive. Nobody has a title and there are no promotions. It has been self-managed since the early 1990s. The biologists who select new varieties of tomato, the farm hands who pick them, the factory workers who process them and the accountants in the office are all equally responsible.

There are not even any budgets: people negotiate expenditure with their colleagues, and decisions are taken by those closest to the place where it will have most impact. Each employee has a 'colleague letter of understanding' rather than a job description or contract of employment. This sets out not just their responsibilities, but their performance indicators. They write this letter themselves, negotiating its content and their pay with their peers based on their performance. The highest-paid employee receives only six times more than the lowest, an unusually small ratio for a fairly large company. The company is famously lacking in the usual politicking about money and status. People feel far more committed to their peers than they ever would to their bosses.

The story of how this came about is as follows. When Chris Rufer, the founder of Morning Star, went into the processing business in 1990, he brought together his employees 'at a small farmhouse on a dirt road on the outskirts of Los Banos, California', writes Paul Green of the Self-Management Institute. He asked the question: 'What kind of company do we want this to be?', and the answer built upon three principles: that people are happiest when they have personal control over their life; that people are 'thinking, energetic, creative and caring'; and that the best human organisations are ones like voluntary bodies that are not managed by others, but in which participants coordinate among themselves. Defying the cynics, the system continued to work as Morning Star grew into a firm with four hundred full-time (and 3,000 part-time) employees.

Far from being a recipe for chaos, self-management works brilliantly. Yet, apart from a few business-school studies, Morning Star's continuing success has been neglected by both the media and the academic world, partly because the firm runs so smoothly and is rarely in the news, partly because food processing is unfashionably low-tech and the company is in California's dull Central Valley, and partly because the ethos that founded it is highly libertarian. Chris Rufer comes to it as a believer in freedom of opportunity, rather than necessarily in equality of outcome. This makes him – in the ludicrous Alice-in-Wonderland world of the media – 'right-wing'. So he does not get put on a pedestal as a great corporate reformer who empowers the workers, even though he is one. Hundreds of firms have come to learn about self-management from Morning Star and gone away enthused. But very few have emulated it, because the initial enthusiasm is lost in a swamp of reports and meetings when they get back to head office. Starting a self-managed business from scratch, as Rufer did, is one thing. Asking employees of an existing firm to lay down their perks is quite another.

Yet, bit by bit and far too slowly, the idea will catch on. Morning Star and others that are trying self-management, such as the online retailer Zappos, are in my view only doing explicitly and enthusiastically what other companies are gradually being forced to do implicitly and reluctantly. The old idea that one lot of workers – the ones who wear suits and speak at conferences – should be 'in charge' of telling the rest, who wear T-shirts and jeans, what to do is bizarre when you think about it. Why not think of the white-collar executives as hired servants working for the productive members of a firm?

Whole Foods, an American food retailer, delegates decisions on what to stock and how to promote its goods to local stores and to teams within the stores. The company also operates a scheme called gainsharing, in which the bonuses earned by each team can be shared with other teams. John Mackey, Whole Foods's co-founder, is a committed supporter of the power of free markets to disrupt and flatten society's inequalities. He is also somebody who sees evolution at work in the market: 'Business is not really a machine, but part of a complex, interdependent and evolving system with multiple constituencies.'

Oh, and do pause to compare Morning Star (with its slightly Soviet name) with the collective farms of Stalin's Russia and Mao's China. Russian and Chinese peasants were forced to join collective farms, refused the chance to leave, given production targets from the centre, told by bosses what work to do and had to watch their produce confiscated to be distributed by the state. Little wonder that many Russians called it a second serfdom. Was there ever a better example of how true egalitarianism comes from liberty, rather than from the state?

The evolution of economic development

Until two hundred years ago nearly all the world was poor. A handful of European and North American countries then escaped to unimaginable comfort, health and opportunity for the majority of their citizens, leaving most of the rest of the world behind. In the past few decades many more countries have followed that path to begin the great escape from poverty, mostly in Asia, while others remain far behind, mostly in Africa. This process of economic development is by far the most momentous and extraordinary thing that has happened in recent decades. Yet there is no 'Great Man' (or woman) who can take the credit. In fact, the closer you look at the history of economic development, the less you find that it has owed to leadership.

Economic development is more than just a growth of income – it is the appearance of a whole system of collaborative engagement among people to drive innovation that cuts the time it takes people to fill needs. And to this day, despite the fact that we know economic development can happen almost everywhere, and we know some of the conditions that make it possible, we still cannot really make it happen to order. A series of papers by the Princeton economist Dani Rodrik and his colleagues tried to shed light on the impact of policy decisions on economic growth, but found that 'most instances of economic reform do not produce growth accelerations', and 'most growth accelerations are not preceded or accompanied by major changes in economic policies, institutional arrangements, political circumstances, or external conditions'. The economist William Easterly points out that the evidence for a change of leadership being the cause of a growth miracle anywhere in the developing world is wholly lacking: the timing simply does not match. The effect of leaders on growth rates, he says, is close to zero, a conclusion that is 'almost too shocking to be believed'.

South Korea and Ghana had the same income per capita in the 1950s. One received far more aid, advice and political intervention than the other. It is now by far the poorer of the two. In general, Asian economies grew their way out of poverty in the late twentieth century, while African economies failed to be aided out of poverty. Trade, not aid, proved the best way to achieve an increase in prosperity. And just when experts were beginning to despair of Africa ever achieving economic development, and sometimes even to reach for racial or institutional explanations of why this should be, Africa suddenly began to experience its own development miracle, which continues to this day: the GDP of many African countries has doubled in a decade. The story of economic development is a bottom–up story. The story of lack of development is a top–down story.

Indeed, the case against creationism in economic development is even stronger than that. The real cause of poverty today – now that it is avoidable – is the unchecked power of the state against poor people without rights, says William Easterly. Implicitly, today's development industry yearns for autocrats advised by experts, and quite often that is what it gets: a tyranny of experts. But such tyrannies of experts all too often become tyrannies more generally, the money and methods of aid being only too helpful to the cause of autocrats. Spontaneous solutions by free individuals would have achieved far more development than has happened. As Deirdre McCloskey argues, 'The importation of socialism into the Third World, even in the relatively non-violent form of Congress-Party Fabian-Gandhism, unintentionally stifled growth, enriched large industrialists, and kept the people poor.'

Easterly's case rests on a detailed analysis of the history of aid, from its beginnings with the Rockefeller Foundation in the 1920s in China, through its post-war expansion under government funding in Africa, Latin America and Asia, to its most recent expression in the great private and public philanthropies of today. He is careful to state – as am I – that humanitarian help is a good thing, and that getting food to famine victims, medicine to disease victims and shelter to disaster victims is absolutely the right thing to do. Aid is vital to the alleviation of crises – such as the Ebola epidemic of 2014–15. The disagreement is around whether aid can remedy poverty, rather than merely respond to crises. Giving money to poor people is not a sustainable solution to poverty. So how do you help poor people? Do you instruct, plan and order their lives with expertise and lots of government, or do you get them freedom to exchange and specialise, so that prosperity can evolve?

Friedrich Hayek and Gunnar Myrdal shared the Nobel Prize in Economics in 1974 for answering this question in opposite ways. Hayek thought that individual rights and freedoms were the means by which societies escaped from poverty. Myrdal thought development would be 'largely ineffective' without 'regulations backed by compulsion', because a 'largely illiterate and apathetic citizenry' would achieve nothing without government direction. Myrdal – correctly – claimed to represent the consensus view on development: 'It is now commonly agreed that an underdeveloped country should have an overall, integrated national plan.' Hayek's approach, by the 1970s, barely existed within Western governments and international agencies. (Bafflingly, the opponent of state compulsion ended up being labelled as 'right-wing'.)

Myrdal's approach had been foreshadowed in the Rockefeller Foundation's attempt to put together an integrated plan to attack rural poverty in China in the 1920s. As Easterly points out, this was essentially a way to change the subject from the occupation of enclaves within China by privileged foreigners. The West looked to turn its occupations into technocratic expertise on development. Chiang Kai-shek, needing funds to support his autocratic ambitions, was only too happy to go along with it. The Rockefeller Foundation supported the Chinese economist H.D. Fong, whose vision of authoritarian development Chiang adopted. Development aid ended up supporting the ambitions of a dictator, whose mistakes in turn opened the way for the tyranny of communism. In that sense, well-intentioned aid money may have played a role in creating the world's most murderous regime. Fong's Rockefeller-funded colleague, the economist John Bell Condliffe, saw what was happening, and warned presciently in 1938: 'We face a new and more formidable superstition than the world has ever known: the myth of the nation-state, whose priests are as intolerant as those of the Inquisition.' Condliffe saw autocratic power as the cause of, not the solution to, poverty.

Much the same happened in post-colonial Africa after the Second World War. Britain withdrew, allowing strong men to take over most countries. But before the British left, they put in place a system of technocratic development that ensured a ready supply of command, control and money for the strong men to appropriate. Why did they do this? Lord Hailey, a retired colonial official, came up with this approach during the Second World War, when the success of Germany and Japan threatened British prestige and made pith-helmeted district commissioners seem less god-like. He argued that the British Empire should portray itself as a 'movement for the betterment of the backward peoples of the world'. It would thus reinvent itself as a progressive force. And of course, this required 'a far greater measure of both initiative and control on the part of the central government'. So Britain's administration in its colonies suddenly became less about administering justice and much more about promoting economic development. This provided an excuse for sidelining questions of independence – until the subject people were 'ready'. Hailey got the Americans to go along with this, by suggesting a similar line on Southern segregation. Economic betterment would come first; political liberation could wait.

The consequence was that the newly liberated 'Third World' of the 1950s and 1960s was handed autocratic assumptions ready-made. 'The masses of the people take their cue from those who are in authority over them,' said the United Nations' 'Primer for Development' in 1951. Hayek was not deceived. He saw the United Nations charter as meaning 'a more or less conscious endeavor to secure the dominance of the white man'.

The very same philosophy of technocratic development came to be extremely useful to the Americans in the Cold War. They could disguise their support for anti-Soviet allies under a covering of neutral aid, distributing World Bank loans in places like Colombia both to promote development and to buttress anti-communist regimes. Once again, aid was used to strengthen autocrats. Part of the problem was that rich governments saw the nation state as the unit of development, rather than the individuals within and between such countries. Authoritarian regimes were discredited by the middle of the twentieth century in Europe and Japan. But they were given a new lease of life in the developing world, where the nation state was effectively buttressed by aid from America and Europe. 'Development lent unintended support for suppressing minority rights in the name of putting the nation's collective wellbeing above all else,' says Easterly.

Easterly is no more forgiving when it comes to modern aid initiatives. Tony Blair's Africa Governance Initiative states as its goal: 'strengthening the government's capacity to deliver programmes'. In Ethiopia this has meant support for the regime's 'villagisation' projects, in which more than a million families are relocated to model villages, freeing land to be sold to foreign investors. Considerable unrest and violence has ensued, yet the programme wins not just funds but praise from international agencies. According to a 2010 report from Human Rights Watch entitled 'How Aid Underwrites Repression in Ethiopia', the Ethiopian leader Meles Zenawi used aid money to blackmail his citizens, denying food relief to hungry people if they supported the opposition.

Another example is Malawi, where European Union development aid to help the country diversify away from growing tobacco and towards growing sugar has had the perverse consequence of encouraging the expropriation of land from smallholders. The aid has effectively provided an incentive for some wealthy people to seek help from the police and village chiefs to evict people from land so they can grow now profitable sugar on larger plots. Predatory elites have been the bane of poor countries in Africa and Latin America for decades, and aid has often – wittingly or not – subsidised those predators.

The evolution of Hong Kong

From ancient Egypt to modern North Korea, always and everywhere, economic planning and control have caused stagnation; from ancient Phoenicia to modern Vietnam, economic liberation has caused prosperity. The paradigmatic example is the city state of Hong Kong, whose history is a shining example of what economic development could be.

Hong Kong's story as a British enclave begins in a disgraceful episode of imperialism, with Britain imposing an addictive narcotic on the Chinese at the point of a gun in the Opium Wars. But after that, more by evolutionary accident than design, Hong Kong became a place of peaceful and voluntary trade, with light-touch government. Sir Harry Pottinger, an Irishman who became Hong Kong's first Governor in 1843, set his face against colonising or ruling even part of China, arguing instead for a free-trading entrepôt. So he refused to tax trade at all; he refused to ban any country from trading there, even Britain's enemies; and he respected local customs. Pottinger was not popular with the British residents, who wanted something much more like conquest and tribute, but he sowed seeds of free trade that gradually flourished. More than a century later, in the 1960s, Sir John Cowperthwaite, the Financial Secretary of Hong Kong, resumed the experiment. He refused all instruction from his LSE-schooled masters in London to plan, regulate and manage the economy of his poor and refugee-overwhelmed island. Let merchants free to do what merchants can, was his philosophy. He rewarded his bureaucrats for coming in under budget, an unusually rare trait in the public sector. He allowed three stock exchanges and undermined the monopoly powers of British businessmen. At London's insistence he politely asked Hong Kong's merchants if they would like to pay income tax, to which the answer was as predictable as it was furious. In short, he tried Adam Smith's recipe. Today Hong Kong has higher per capita income than Britain.

13

The Evolution of Government

Since as men clawed to the pinnacle of office, all the time
They strewed their path with perils. And at the apex of their climb,
Often Envy would blast them like a thunderbolt, to fell
Them with disdain and hurl them in the pit of hateful Hell

Lucretius, *De Rerum Natura*, Book 5, lines 1123–6

To judge by the movies, in the American west in the nineteenth century, homicide was routine. Cattle towns like Abilene, Wichita and Dodge City were places where the lack of government – if there was government at all, it was in the form of a timid, corrupt or outgunned sheriff – resulted in endless Hobbesian slaughter. Was this really the case? Actually, in five such cattle towns in the key years 1870–85, there was an average of just 1.5 murders per town per cattle-trading season. That's a lower murder rate than today in that part of America, let alone in its big cities. Yet if anything, the population of the cattle towns was higher in those times. Wichita alone experiences up to forty murders a year today, with the full might of the state and federal authorities in charge.

The truth is that the wild west was without much government, but it was very far from lawless, or even violent. As the economists Terry Anderson and P.J. Hill document in their book

The Not So Wild, Wild West, with few formal law-enforcement mechanisms, people generated their own arrangements, enforced by private marshals and punished by simple measures such as exile from a wagon train. Anderson and Hill conclude that in the absence of a government monopoly of coercion, multiple private law enforcers emerged, and competition among them drove improvements and innovations that thrived by natural selection. In effect, the cattlemen of the nineteenth century rediscovered what medieval merchants had found – that customs and laws would emerge where they were not imposed. It was very far from anarchic.

Robert Ellickson of Yale documented a good example of this more recently in Shasta County, California, an area of farms and ranches. Taking his cue from a famous example given by the economist Ronald Coase (who argued that in the absence of transaction costs, wrongs between cattle ranchers and wheat farmers would be righted by private negotiation rather than state punishment), Ellickson looked to see how individuals actually dealt with trespassing cattle. He found that the law was largely irrelevant. People dealt with the problem privately, sometimes even illegally. For example, they would call the owner of the cattle and ask him to retrieve his errant beasts; if he failed to do so persistently, he would be punished by finding his animals driven away in the wrong direction, or even castrated. Everybody knew he stood a good chance of one day being on the wrong end of a complaint, so was keen to reciprocate apologetic responses. This is just a rural version of good neighbourliness. Somebody who resorts too quickly to the police, or the courts, to deal with a problem neighbour is generally thought to have behaved badly and to forfeit the community's good will.

Government at its root is an arrangement among citizens to enforce public order. It emerges spontaneously at least as much as, perhaps more than, it is imposed by outsiders. And over the centuries it has changed form organically, with very little planning.

The evolution of government in prison

In a fascinating recent study of prison gangs, entitled *The Social Order of the Underworld*, David Skarbek finds evidence that they too are examples of the emergence and elaboration of spontaneous order, albeit backed by the threat of violence. American prisons have never been wholly reliant on the state for order. The governor and the guards are there, sure, but most of the 'law' is a spontaneously emergent habit among prisoners, known as the 'convict code'. This takes the form mainly of honour among thieves, its fundamental premise being, in the words of Donald Clemmer, who conducted the seminal study of norms in jail: 'Inmates are to refrain from helping prison or government officials in matters of discipline, and should never give them information of any kind, and especially the kind which may work to harm a fellow prisoner.' Skarbek points out that this code evolved, rather than was invented. No group of inmates met to decide it. Although transgressors were punished with ostracism, ridicule, assault or death, punishment was decentralised. Nobody was in charge. And the convict code 'facilitated social cooperation and diminished social conflict. It helped establish order and promote illicit trade.'

However, in the 1970s the convict code began to break down in male, though not in female, prisons. This coincided with a rapid increase in the prison population, and more heterogeneous ethnic diversity among inmates. This fits what we know about pre-state societies. When villages or bands get beyond a certain size, interpersonal codes of conduct become unworkable. There is too much anonymity. Violence increased markedly, but something else began to happen too: prison gangs began to emerge.

Throughout America's prison system, mainly in the 1970s, gangs began to appear. They had little or nothing to do with gangs in the outside world, and arose in regions with no such street gangs. They emerged in thirty different prisons. It was as if somebody had imposed the idea of gangs as a kind of reform. Yet not only did the gang culture emerge from the inmates rather than from the officials, it did so unconsciously; and although there are gang leaders, the system as a whole is highly decentralised. As Skarbek puts it, 'the social order that exists was not chosen. No one is in charge.' Echoing the Scottish philosopher Adam Ferguson, Skarbek concludes: 'This bottom–up process of institutional emergence was the result of inmate actions, but not the execution of any inmate design.' It evolved.

The Mexican mafia in San Quentin was the first such gang, and remains one of the most powerful, but others soon followed. The effect of the gangs was to suppress violence, increase trade in drugs and other goods, lower prices, and generally improve the inmates' lives. Skarbek analyses how this happened, and rules out all explanations of the phenomenon but one: that what was happening was the emergence of a rudimentary form of government. The appearance of gangs was a solution to the lack of governance among convicts. The prison officials generally welcomed the gangs, knowing that they helped to maintain order. And the reason gangs did not form in female prisons was simply because their population was still small enough for norms, codes of conduct, to work instead. In other words, government begins as a protection racket and emerges spontaneously when population reaches a certain size. The Mexican mafia now controls the Californian drug trade not just in prisons but on the street, extracting rent from drug dealers and enforcing its power by the threat of violence within the prisons. One of the reasons for the recent decline of violence in the United States may be that gangs have managed to impose slightly more order on the drug trade.

The evolution of protection rackets into governments

So if gangsters become governments, does this mean that governments began as gangsters? As Kevin Williamson argues in his book *The End is Near and it's Going to be Awesome*, organised crime and government are more than first cousins; they are sprung from the same root. That is to say, government began as a mafia protection racket claiming a monopoly on violence and extracting a rent (tax) in return for protecting its citizens from depredation by outsiders. This is the origin of almost all government, and today's mafia protection rackets are all in the process of evolving into government. The Mafia itself emerged in Sicily in a time of lawlessness when property rights were insecure and plentiful ex-soldiers were prepared to offer their services as paid protectors. The Russian mafia emerged in the 1990s in a similar way: a lawless time, a lot of ex-soldiers looking for work.

Throughout history, the characteristic feature of the nation state is its monopoly of violence. In ancient Rome, especially during the first century BC, consuls, generals, governors and senators, each with his own organised crime syndicate of thugs and legions, fought over the division of the spoils of imperial conquest in a series of civil wars, assassinations and plots that grew steadily more desperate – until one emerged with sufficient wealth and power to impose a monopoly of military might. He called himself Augustus and ushered in a *Pax Romana* that lasted, with occasional bloody interruption, for two centuries. As Ian Morris argues in his book *War: What is it Good For?*, 'the paradoxical logic of violence was at work. Because everyone knew that the emperor could (and if pressed would) send in the legions, he hardly ever had to do so.'

Today we generally take a benign view of the state as an institution that tries to be fair and just, and that exists to tame the worst instincts of individuals. But think about the history of this institution. Virtually everywhere – the United States and some other ex-colonies being notable exceptions – government originated as a group of thugs who, as Pope Gregory VII trenchantly put it in the eleventh century, 'raised themselves above their fellows by pride, plunder, treachery, murder – in short by every kind of crime'. For most of history, the state has been an 'ever-present predator and all-around abuser of human rights', in the economic historian Robert Higgs's words. George Washington said that 'Government is not reason. It is not eloquence. Government is force. Like fire, it is a dangerous servant, and a fearful master.' The social critic Albert Jay Nock, writing in 1939, had become especially cynical, with good reason: 'The idea that the State originated to serve any kind of social purpose is completely unhistorical. It originated in conquest and confiscation – that is to say, in crime.' Perhaps we have left all that behind, and the state is now evolving steadily towards benign and gentle virtue. Perhaps not.

Tudor monarchs and the Taleban are cut from exactly the same cloth. Just as Henry VII acted like a Corleone, so Islamic State, the Colombian FARC, the Mafia itself, the Irish Republican Army all come to behave more and more like government – enforcing a strict moral code, 'taxing' commodities (opium, cocaine, waste disposal), punishing transgressors, providing welfare. And even modern governments have an element of the crime syndicate about them. Police forces repeatedly harbour criminals all over the world: the US Department of Homeland Security is only a little more than a decade old, but in 2011 over three hundred of its employees were arrested for crimes such as drug smuggling, child pornography and selling intelligence to drug cartels.

Like Augustus's legions, the state's monopoly of weaponry stays out of sight as much as possible. But it's there. Many people are bothered about the number of privately owned guns in the United States, but what about publicly owned ones? In recent years the United States government (not the military) has purchased 1.6 billion rounds of ammunition, enough to shoot the entire population five times over. The Social Security Administration ordered 174,000 rounds of hollow-point bullets. The Internal Revenue Service, the Department of Education, the Bureau of Land Management, even the National Oceanographic and Atmospheric Administration, all have guns.

When riots broke out in Ferguson, a suburb of St Louis,

Missouri, in August 2014, many people were shocked that the police appeared in armoured vehicles with mounted weapons and wearing uniforms and gear that looked more military than law-enforcement. Senator Rand Paul commented in *Time* magazine that the federal government had incentivised the militarisation of local police, funding municipal governments to 'build what are essentially small armies'. Evan Bernick of the Heritage Foundation had warned a year earlier that the Department of Homeland Security had handed out anti-terrorism grants to towns across the country so they could buy armoured vehicles, guns, armour, and even aircraft. Indeed, the Pentagon actually donates military equipment to the police, including tanks. Radley Balko, a reporter for the *Washington Post*, has chronicled the blurring of the line between police and military inherent in the 'wars' on drugs, poverty and terror. The police have come to resemble an occupying army who see the citizenry as the enemy. Senator Paul thinks the militarisation of law enforcement combines with an erosion of civil liberties to create a very serious problem. Yet the truth is that this is not so much a new problem as an old problem that would have seemed all too familiar to the founding fathers of America, confronted with redcoat regiments marching through their streets.

The libertarian Levellers

So government began as a protection racket. Until about 1850 it was taken for granted that a liberal and progressive person would be mistrustful of government. From Lao Tzu, castigating the dictatorial *dirigisme* of the Confucian state, with its 'laws and regulations more numerous than the hairs of an ox', to the *sans-culottes* of 1789, those who wanted to improve the lot of the poor saw government as the enemy. Government was something that lived parasitically off the backs of the working people, spending the money it extorted on war and luxury and oppression. 'The danger is not that a particular class is unfit to govern,' said Lord

Acton. 'Every class is unfit to govern.' The problem is not the abuse of power, echoed the motivational speaker Michael Cloud more recently, but the power to abuse.

The church at Burford in Oxfordshire is a shrine of pilgrimage for people from the radical left. It was here in 1649 that Oliver Cromwell imprisoned three hundred mutineering Levellers and had three shot for refusing to recant. Most people today think of the Levellers as being like the Diggers, that is to say socialists *avant la lettre* – egalitarian, communitarian, revolutionary. Yet, as Daniel Hannan and Douglas Carswell, free-market MEP and MP respectively, argue, this misreads history. The Levellers were what we would today call libertarians or classical liberals. They argued for private property, free trade, low taxes, limited government and freedom of the individual. The enemy for them was not commerce, but government. They had taken part in a rebellion, beheaded a king, and were frustrated now that a corrupt and complacent Parliament refused to hold fresh elections, and refused to guarantee the ancient economic freedoms they felt were their birthright. Meanwhile their general seemed more and more to see himself as a Messianic figure chosen by Providence to rule as a tyrant. Their immediate beef with Cromwell was that they did not want to pursue his religious and ethnic crusade against the Irish, but their libertarianism was political, economic and personal.

In their manifesto of 1649, 'An Agreement of the Free People of England', the four leaders of the movement, John Lilburne, Thomas Walwyn, Thomas Prince and Richard Overton demanded, from prison in the Tower of London, that politicians be restrained from raising too much tax or restricting too much trade, sentiments rarely heard on the left today:

> That it shall not be in their power to continue to make any Laws to abridge or hinder any person or persons, from trading or merchandising into any place beyond the Seas, where any of this Nation are free to trade.

Little wonder that the Levellers have received the approbation of modern free marketeers, from Friedrich Hayek and Murray Rothbard to Hannan and Carswell.

Commerce as the midwife of freedom

By the end of the seventeenth century, European states had invented centralised, bureaucratic government whose chief job was to maintain order – Thomas Hobbes's Leviathan. Then came the Glorious (English), American and French revolutions, and the idea that government should be tamed, liberated, limited and held accountable to 'the people'.

Until 1850 nobody would have batted an eyelid at the equation between, on the one hand, free trade, limited government and low taxes, and on the other, championing of the poor and relief of the needy. Throughout the eighteenth century, the champions of *laissez-faire* – the people who thought free exchange of goods and services was the best way to improve the general wellbeing – were on the political 'left'. The Whigs of 1688, the rebels of 1776, and the thinkers who inspired them, from Locke and Voltaire to Condorcet and Smith, were radical progressives and free-market, small-government liberals. (Voltaire made a fortune as a grain trader.) It would have made no sense to have argued that the state was the organ of liberty and progress. These were, remember, the days when the state not only claimed a monopoly of violence and the power to decide what might be traded, but prescribed in intrusive detail your religious observance, censored your speech and writing, and even mandated your dress according to class. Not only that, but as Stephen Davies points out, a new eighteenth-century idea, especially in Germany, was taking hold – that of the 'police state', which meant that every citizen was a servant of the state. Frederick the Great called himself the first servant of the state, with the emphasis as much on 'servant' as on 'first'. So the radicals who espoused freedom to exchange goods and services also espoused freedom of thought and action.

As an illustration of just how radical an idea the free market was, in 1793, in Edinburgh – the supposedly enlightened Athens of the North – one Thomas Muir was tried for sedition, the prosecution alleging that he had scandalously argued that 'taxes would be less if they were more equally represented'. He got fourteen years' transportation to Australia. William Skirving and Maurice Margarot got the same sentence for echoing Adam Smith on free trade. Little wonder that the next year Dugald Stewart, later the biographer of Adam Smith, decided to apologise abjectly for even mentioning Condorcet's name in a book. Enlightenment had to hide under a bushel.

Free trade and free thinking

Contrast the philosophies of Thomas Jefferson and Alexander Hamilton. Jefferson, the gent, had imbibed the philosophy of the Enlightenment and worshipped at the shrine of Lucretius. But in the end he wanted an agrarian, protected, hierarchical, stable Virginian society. He hated the way people lived 'piled upon one another in large cities', and he suggested that America 'let our workshops remain in Europe'. It was Hamilton, the immigrant, living in chaotic Manhattan, who embraced the future – the creative destruction that commerce and abundant capital would bring, the dissolving of social strata, the upending of power (although he did argue for a small tariff to protect infant industries).

In Britain, the founders of the anti-slavery society were free traders. Read, for example, the writings of Harriet Martineau, who shot to fame in the 1830s because of her series of short fictional books called *Illustrations of Political Economy*. These were intended to educate people in the ideas of Adam Smith ('whose excellence is marvellous', she said) and other economists. They are all about the virtues of markets and individualism. Today, most people would call her right-wing. Yet Martineau was a firebrand feminist, a working woman who lived by her pen, and a political radical who her contemporaries saw as almost dangerous (Charles Darwin's father grew worried when she befriended his two respectable sons). She toured America, speaking passionately against slavery, and became so notorious that in South Carolina there were plans to lynch her. But there is no contradiction: her economic libertarianism was part and parcel of her political libertarianism. Liberals were trying to lift the dead hand of the corrupt and tyrannical state from the market economy as well as from the private life of the citizen. In those days, to be suspicious of a strong state was to be left-wing.

In early-nineteenth-century Britain, free trade, small government and individual autonomy went together almost automatically with opposition to slavery, colonialism, political patronage, and an established Church. The mob that surrounded King George III's carriage as he went to open Parliament in 1795 was demanding free trade in corn and the lifting of multiple and detailed regulations about the sale of bread. The rioters who broke into Lord Castlereagh's house in 1815 were against protectionism. The peaceful demonstration in Manchester that was charged by cavalry in 1819 – the 'Peterloo massacre' – was in favour of free trade as well as political reform. The Chartists who spearheaded working-class consciousness were founding members of the Anti-Corn Law League.

Or take Richard Cobden, the great champion of free trade responsible more than anybody else for that extraordinary spell between 1840 and 1865 when Britain set the world an example and unilaterally and forcefully dismantled the tariffs that entangled the globe. (Cobden comes close to being a Great Man.) He was a passionate pacifist, prepared to make himself unpopular for opposing the Opium War and the Crimean War, deeply committed to the cause of the poor, heckled as a dangerous radical when he first spoke in the House of Commons, and independent enough to refuse to serve as a government minister under two prime ministers, and to refuse a baronetcy from a monarch he disapproved of. He was a genuine radical. Yet he embraced free trade as the best possible means for achieving both peace and prosperity for all. 'Peace will come to earth when the people have more to do with each other and governments less,' he said, sounding like a member of the Tea Party. So pure was his support for free trade that Cobden even lambasted John Stuart Mill for briefly flirting with the idea that infant industries needed protection in their early years. He took the ideas of Adam Smith and David Ricardo and implemented them. The result was an acceleration of economic growth all around the world.

There it is again, the peaceful co-existence in one head of causes embraced by today's left and today's right. Political liberation and economic liberation went hand in hand. Small government was a radical, progressive proposition. Between 1660 and 1846, in a vain attempt to control food prices by prescription, the British government had enacted an astonishing 127 Corn Laws, imposing not just tariffs, but rules about storage, sale, import, export and quality of grain and bread. In 1815, to protect landowners as grain prices fell from Napoleonic wartime highs, it had banned the import of all grain if the price fell below eighty shillings a quarter (twenty-eight pounds). This led to an impassioned pamphlet from the young theorist of free trade David Ricardo, but in vain (his friend and supporter of the Corn Law, Robert Malthus, was more persuasive). It was not until the 1840s, when the railways and the penny post enabled Cobden and John Bright to stir up a mass campaign against the laws on behalf of the working class, that the tide turned. With the famine in Ireland in 1845, even the Tory leader Robert Peel had to admit defeat.

Cobden's astonishing campaign against the Corn Laws, then against tariff protection more generally, succeeded eventually in persuading not just much of the country, and most intellectuals, but the leading politicians of the day, especially William Ewart Gladstone. The great reforming chancellor and prime minister championed all sorts of progressive causes, from the plight of the poor to home rule for Ireland, and in economics he was a convinced free trader, who steadily shrank the size of the state. In the end Cobden and his allies even won over the French. Cobden persuaded Napoleon III of the virtues of free trade, and himself negotiated the first international free-trade treaty in 1860, the so-called Cobden–Chevalier Treaty. That treaty also established the principle of unconditional 'most-favoured nation' clauses, and led to a cascade of tariff-dismantling all over Europe, in effect creating a giant free-trade area for the first time in modern history, though of course not all goods were affected. Italy, Switzerland, Norway, Spain, Austria and the Hanseatic cities quickly followed suit and disarmed their tariffs.

What Adrian Wooldridge and John Micklethwait, in their book *The Fourth Revolution*, call the liberal state may have begun with John Locke, been championed by Thomas Jefferson, found its clearest exponent in John Stuart Mill, and reached its most radical extreme with Richard Cobden, but with the benefit of hindsight we can see that it was not invented by anybody. It emerged; it evolved.

The counter-revolution of government

Yet Cobden's achievements began to erode as the nineteenth century wore on. In the late 1870s Bismarck's Germany suffered from an overvalued currency, which resulted in a recession. The cause was the huge capital inflow of five billion francs from France in payment of the war indemnity that the French had been forced to pay after the Franco–Prussian War to regain captured territories. In response to this recession, and the election of a more conservative parliament following an attempted assassination of the Kaiser, in 1879 Bismarck brought in the 'iron and rye' tariff to protect German industry and agriculture. It was the start of a long succession of competing tariff increases from 1880 until the start of the First World War, in America, France and South America. Only Britain stood alone, defiantly refusing to introduce tariffs, or retaliate against those who did, until well into the twentieth century. Despite strong pressure from Joseph Chamberlain and his Tory allies for 'tariff reform' and 'imperial preference', Britain's almost religious devotion to free trade persisted up until and after the First World War. Then gradually the Liberal Party was squeezed out by imperial-preference Conservatives from the right, and protectionist Labour candidates urging self-sufficiency from the left. Still, it was not until 1932 that Neville Chamberlain brought in a general tariff.

The return of protectionism was part of what Brink Lindsey has called the industrial counter-revolution that began in the last quarter of the nineteenth century – when suddenly progressives and radicals decided that the state was no longer their enemy but their friend. A new alliance was born between nostalgic reactionary conservatives, who wanted hierarchy preserved amid the dizzying ferment of innovation unleashed by the Industrial Revolution, and progressive reformers who thought government should lead social change. As Deirdre McCloskey diagnosed: 'The sons of bourgeois fathers became enchanted . . . by the revival of secularised faith called nationalism and of secularised hope called socialism.' You can see this in Karl Marx and Friedrich Engels, with their horror of economic change. 'Constant revolutionizing of production, uninterrupted disturbance of all social conditions, everlasting uncertainty and agitation distinguish the bourgeois epoch,' Marx and Engels wailed in *The Communist Manifesto*, and '. . . all that is solid melts into air, all that is holy is profaned'. Or take William Morris and his fellow socialists, mourning the loss of stable, simple medieval Merrie England, building the new socialist Jerusalem on a fantasy of Arthurian legends.

In the arts you can detect the shift quite clearly. In the early part of the nineteenth century many poets, novelists and playwrights were ardent supporters of classical liberalism, free trade and limited government. See the works of Schiller, Goethe and Byron. Giuseppe Verdi's *Rigoletto* and *Aida* contain very liberal narratives about the nature of power. The open commercial society had liberated artists from the system of patronage, as they were able to sell their works in a mass market rather than rely on a wealthy individual. However, as time went on, many artists became hostile to liberalism, seeing bourgeois society as stultifying. The critics of the liberal order included Henrik Ibsen, Gustave Flaubert and Emile Zola. These opponents played an important role in portraying the liberal order in a negative light.

The true radicals, the people with a vision of liberty and change, people like Cobden and Mill and Herbert Spencer, were then quite unfairly dumped on the 'right'. Nobody would have thought them right-wing in their time – they were pacifists, egalitarians, feminists, liberals, internationalists, religious free thinkers. But their affection for free markets as the best way of achieving these goals catapulted them, in twentieth-century eyes, all the way across the political spectrum from left to right.

All those centuries of struggling against the power of monarchs and their henchmen are suddenly forgotten when there is a chance of appointing the henchmen yourselves. No longer is safeguarding individual liberty the chief purpose of politics; from now on there is to be planning and welfare. The revolution will henceforth be a top–down affair, directed by the enlightened leaders of the proletariat. Liberalism had learned to 'place no small confidence in the beneficent effects of the central state', wrote A.V. Dicey in 1905.

Business, too, embraced government intervention. As the nineteenth century ended robber-baron industrialists were only too eager to rush off to form cartels, or welcome government regulations, the better to extinguish wasteful competition. Yet instead of earning the ridicule of the economics profession for this cronyism – as they had done from Adam Smith – they were now applauded. Thought leaders of the left, like Edward Bellamy and Thorstein Veblen, demanded an end to duplication and fragmentation in business. There must be a plan, a planner and a single structure, they agreed. Bellamy's vision of the future, in his immensely influential and bestselling novel *Looking Backward*, has everybody in the future working for a Great Trust and shopping at identical, government-owned stores for identical goods.

Even Lenin and Stalin now admired the big American corporations, with their scientific management, planned workforce accommodation and giant capital requirements. 'We must organize in Russia the study and teaching of the Taylor system and systematically try it out and adapt it to our purposes,' wrote Lenin of the great apostle of scientific management, Frederick Winslow Taylor. The libertarian editor of the *Nation*, Ed Godkin, lamented in 1900: 'Only a remnant, old men for the most part, still uphold the liberal doctrine, and when they are gone, it will have no champions.' The very word 'liberal' changed its meaning, especially in the United States. 'As a supreme, if unintended, compliment, the enemies of the system of private enterprise have thought it wise to appropriate the label,' said Joseph Schumpeter. Everybody, especially on the left, thought the key to the future was command and control, not evolution.

Government was to be the tool by which to engineer society. Around 1900 this was true whether you were a communist wishing to bring in the dictatorship of the proletariat, a militarist wishing to conquer your enemies and regiment your society, or a capitalist wishing to build new factories and sell your products. Once again, this notion of the role of government as planner had not been invented; it had emerged.

Liberal fascism

It is often forgotten that America became a remarkably illiberal place under Woodrow Wilson and his successors. It was not just the tightening of segregation, the spread of eugenic laws and the prohibition of alcohol, but censorship and the squeezing of civil liberties. Jonah Goldberg reminds us that during the First World War a Hollywood producer received a ten-year jail sentence for depicting British troops committing atrocities during the American Revolution.

Some of the rhetoric of Franklin Roosevelt's New Deal had echoes of what was happening in Germany and Italy, and there is abundant evidence that the New Dealers were keen to emulate the apparent success of totalitarian regimes in improving the economy and social order, even if they never contemplated emulating their violence. Plan, plan, plan was the cry on all sides. Joseph Schumpeter thought Franklin Roosevelt was intent on becoming a dictator.

Jonah Goldberg points out in his book *Liberal Fascism* that in the 1930s fascism was widely seen as a progressive movement, and was supported by many on the left: 'Fascism, properly understood, is not a phenomenon of the right at all. Instead, it is, and always has been, a phenomenon of the left. This fact – an inconvenient truth if there ever was one – is obscured in our time by the equally mistaken belief that fascism and communism are opposites. In reality, they are closely related, historical competitors for the same constituents.' Father Charles Coughlin, the 'radio priest' of the 1930s who came closest to imitating Hitler's aims and methods in American politics, was very much a man of the left: criticising bankers, demanding the nationalisation of industry and the protection of the rights of labour. Only his anti-Semitism could be described as 'right'. The very phrase 'liberal fascist' was coined approvingly by H.G. Wells in a speech in Oxford in 1932. Earlier, in 1927, Wells had mused about 'the good there is in these Fascists. There is something brave and well-meaning about them.'

From the perspective of today, or from that of a Cobden-Mill-Smith liberal, there is not a great deal of difference between the various -isms of the twentieth century. Communism, fascism, nationalism, corporatism, protectionism, Taylorism, *dirigisme* – they are all centralising systems with planning at their heart. Little wonder that Mussolini began as a communist, Hitler as a socialist, and Oswald Mosley became a Labour MP very soon after being elected as a Conservative and before turning fascist. Fascism and communism were and are religions of the state.

They are a form of intelligent design. They worship at the feet of a political leader in exactly the way that religions worship at the feet of a god, claiming for him at least some tendency to omnipotence, omniscience and infallibility. In communism there is usually an initial pretence that the leader is not a person but a party, and that the god is a long-dead bloke with a long beard, but it never lasts long. Soon the name of the leader displaces that of Marx: Stalin, Mao, Castro, Kim. True, fascists did not collectivise farms and did allow private companies to operate for profit, but only within state-defined areas and with state-mandated goals. 'Everything in the state, nothing outside the state,' said Mussolini. As Goldberg points out, Hitler hated communists not because of their economic doctrines, or because they wanted to destroy the bourgeoisie – he liked those notions. He championed trade unions in *Mein Kampf*, and attacked the greed and 'short-sighted narrow-mindedness' of businessmen as fervently as any modern anti-capitalist. No, he hated communism because he thought it was a foreign, Jewish conspiracy, as he made clear throughout *Mein Kampf*.

The libertarian revival

The Second World War saw the command-and-control state reach its culmination. Not only were most countries run on strictly authoritarian lines, by fascist, communist or colonial regimes, but even the handful of exceptions, where democracy survived, effectively adopted comprehensive central planning as an emergency measure to fight the war. Certainly in Britain, and to some extent in America, almost every aspect of life was determined by the state. Old-fashioned individualism, or liberalism, was virtually extinct. Or was it? Stirring beneath the wartime centralism, you can detect a handful of voices demanding that when the war is over, we must dismantle the planned economy. People like Herbert Agar and Colm Brogan. The latter warned, in his 1943 book *Who are 'the People'?*: 'Having escaped invasion, the people of Britain have escaped the final test, but ideas are not entirely defeated by the Channel. There is growing support for the theory that the new economic order which the Germans seek to impose will come because it must.'

Most powerful were the voices of refugees from both Hitler and Stalin, who would insist to their Western hosts that Nazi and communist totalitarianisms were not at opposite ends of the spectrum, but were close neighbours: people like Hannah Arendt, Isaiah Berlin, Michael Polanyi and Karl Popper. The most famous of these voices was Friedrich Hayek's, with his prescient warning in *The Road to Serfdom* (1944) that socialism and fascism were not really opposites, but had 'fundamental similarity of methods and ideas', that economic planning and state control were at the top of an illiberal slope that led to tyranny, oppression and serfdom, and that the individualism of free markets was the true road to liberation.

Ignoring Hayek, within months of victory Britain embarked on a comprehensive nationalisation of the means of production in industry, in health, in education and in society. There were few politicians prepared to resist. Even the returned Conservative government of Winston Churchill in 1951 would have continued to mandate identity cards for citizens, had it not been for a libertarian radical named Sir Ernest Benn, a fervent admirer of Herbert Spencer and Richard Cobden, who managed to get the things abolished.

Germany was luckier. In July 1948 Ludwig Erhard, director of West Germany's Economic Council, abolished food rationing and ended all price controls on his own initiative, trusting to the market. General Lucius Clay, military governor of the American zone of occupation, called him and said, 'My advisers tell me what you have done is a terrible mistake. What do you say to that?' Erhard replied: 'Herr General, pay no attention to them! My advisers tell me the same thing.' The German economic miracle was born that day; Britain kept rationing for six more years.

Government as God

Yet creationism in government shows no sign of fading. To this day, despite the resurgence of liberal values that came after the Second World War, and especially after the Cold War, the knee-jerk assumption on the part of much of the intelligentsia is still based on planning rather than evolutionary unfolding. Though politicians are regarded as scum, government as a machine is held to be almost infallible. In the United States, government spending rose from 7.5 per cent of GDP in 1913, to 27 per cent in 1960, to 30 per cent in 2000, to 41 per cent in 2011. The counter-revolution of Ronald Reagan was a mere pause in the advance of government, which has become the conduit of welfare not just from the wealthy to the disadvantaged, but from the middle classes back to the middle classes. Many think government has now evolved to its maximum possible size, that it cannot be sustained on a larger scale.

But the next phase of government evolution is international. The growth of international bureaucracies with power to determine many aspects of people's lives is a dominant feature of our age. Even the European Union is increasingly powerless, as it merely transmits to its member states rules set at higher levels. Food standards, for example, are decided by a United Nations body called the Codex Alimentarius. The rules of the banking industry are set by a committee based in Basel in Switzerland. Financial regulation is set by the Financial Stability Board in Paris. I bet you have not heard of the World Forum for the Harmonisation of Vehicle Regulations, a subsidiary of the UN.

Even the weather is to be controlled by Leviathan in the future. In an interview in 2012, Christiana Figueres, head of the United Nations Framework Convention on Climate Change, said she and her colleagues were inspiring government, private sector and civil society to make the biggest transformation that they have ever undertaken: 'The Industrial Revolution was also a transformation, but it wasn't a guided transformation from a centralized policy perspective. This is a centralized trans-formation.'

Yet perhaps other evolutionary forces are stirring. For years the services which government specialises in providing – care, education, regulation – have been the ones least affected by automation and digital transformation. That may be changing. In 2011 the British government hired a digital entrepreneur named Mike Bracken, and asked him to reform the way big IT contracts were managed. With the support of a minister, Francis Maude, he came up with a system that replaced what he called 'waterfall' projects, which specified their needs in advance and ended up running over budget and out of time, with something much more Darwinian: projects were told to start small, fail fast, get feedback from users early, and evolve as they went along.

When I interviewed Mr Bracken about this approach, which by 2014 had begun to have some striking successes, not least in the gradual but accelerating roll-out of a single government web portal named gov.uk to replace 1,800 separate websites, I realised that what he was describing was evolution, as opposed to creationism. In his 2011 book *Adapt*, Tim Harford had pointed out that whether pacifying Iraq, designing an aircraft or writing a Broadway musical, successful executives had allowed for plenty of low-cost trial and error and incremental change. From the world economy to the laser printer, everything we use comes about by small steps, not grand plans.

The elite gets things wrong, says Douglas Carswell in *The End of Politics and the Birth of iDemocracy*, 'because they end-lessly seek to govern by design a world that is best organized spontaneously from below'. Public policy failures stem from planners' excessive faith in deliberate design. 'They consistently underrate the merits of spontaneous, organic arrangements, and fail to recognize that the best plan is often not to have one.'

14

The Evolution of Religion

Then when the whole earth moves beneath our feet, and cities tumble
To the ground, hit hard, or cities badly shaken, threaten to crumble,
Is it surprising mortal men are suddenly made humble,
And are ready to believe in the awesome might and wondrous force
Of gods, the powers at the rudder of the universe?

Lucretius, *De Rerum Natura*, Book 5, lines 1236–40

On the ceiling of the Sistine Chapel, Adam and God touch fingers. To the uneducated eye it is not clear who is creating whom.
We are supposed to assume God's the one doing the creating,
and much of the world thinks so. To anybody who has read the
history of the ancient world, it is crystal clear by contrast that,
in the words of the title of Selina O'Grady's book on the subject,
Man Created God. God is plainly an invention of the human
imagination, whether in the form of Jahweh, Christ, Allah,
Vishnu, Zeus or Anygod else. The religious impulse is not confined to conventional religion. It animates ghosts, horoscopes,
ouija boards and Gaia; it explains all forms of superstition, from
biodynamic farming to conspiracy theories to alien abduction to
hero worship. It is the expression of what Daniel Dennett calls
the intentional stance, the human instinct to see purpose and

agency and power in every nook or cranny of the world. 'We find human faces in the moon, armies in clouds . . . and ascribe malice and goodwill to every thing that hurts or pleases us,' wrote David Hume in his *Natural History of Religion*.

The urge to impute the shape of every leaf and the time of every death to the whim of an omnipotent deity may seem to be as top–down as it gets. Yet my argument will be that this phenomenon can only be explained as an instance of cultural evolution: that all gods and all superstitions emerge from within human minds, and go through characteristic but unplanned transformations as history unfolds. Thus even the most top–down feature of human culture is actually a bottom–up, emergent phenomenon.

O'Grady vividly tells the story of how Christianity emerged in the first century AD from among a bewildering ferment of competing religious enthusiasms within the Roman empire, and was far from being the most obvious candidate to win global power. The 'single market' of Rome was ripe for a religious monopoly. Empires usually do become dominated by one religion to a great extent: that of Zeus in Greece, Zoroaster in Persia, Confucius in China, Buddha in the Mauryan empire, Mohamed in Arabia.

In first-century Rome, every city had scores of cults and mystery religions competing alongside each other, usually without much jealousy – only the god of the Jews refused to tolerate others. Temples to Jupiter and Baal, Atagartis and Cybele, lay one beside another. Consolidation was inevitable: just as thousands of independently owned cafés were replaced by two or three mighty chains such as Starbucks, with superior products more slickly delivered, so it was inevitable that religious chains would take over the Roman empire. Augustus did his best to pose as a god himself, but that cut little ice with the merchants of Alexandria or the peasants of Asia Minor.

In the middle of the first century, the cult of Apollonius of Tyana looked a better bet to conquer the empire. Like Jesus, Apollonius (who was younger, but overlapped) raised the dead, worked miracles, exorcised demons, preached charity, died and rose again, at least in spiritual form. Unlike Jesus, Apollonius was a famous Pythagorean intellectual known throughout the Near East. His birth had been foretold, he abjured sex, drank no wine and wore no animal skins. He was altogether more sophisticated than the Palestinian carpenter. He moved in grand circles: the dead person he raised was the child of a senator. His fame spread well beyond the Roman lands. When he arrived at Babylon, the Parthian King Vardanes greeted him as a celebrity and invited him to stay and teach for a year. He then travelled east to what is now Afghanistan and India, never to re-emerge. Long after his disappearance his cult competed with the Jewish, Zoroastrian and Christian creeds. Yet eventually it petered out.

Blame Saul of Tarsus, also known as St Paul. Whereas Apollonius had a plodding Greek chronicler as his evangelist, named Philostratus, Jesus was blessed with a peculiarly persuasive if rather eccentric Pharisee who set out to reinvent and convert the Jesus cult into a universal, rather than a Jewish, faith designed to appeal to Greeks and Romans. And St Paul was acute enough to realise that the Jesus cult could be aimed at the poor and dispossessed. Its strictures against wealth, power and polygamy were well designed to appeal to those who had little to lose. Quite how the Christians eventually (three centuries on) persuaded an emperor, Constantine, to convert to their cause remains a little mysterious, but it surely had much to do with the populist appeal of the new creed. After that, the conquest of large swathes of the planet by the Christian religion owed as much to power as to persuasion. All competing religions were ruthlessly and violently stamped out wherever possible, starting with the Emperor Theodosius.

In short, you can tell the story of the rise of Christianity without any reference to divine assistance. It was a movement like any other, a man-made cult, a cultural contagion passed from mind to mind, a natural example of cultural evolution.

The predictability of gods

Further evidence for the man-made nature of gods comes from their evolutionary history. It is a little-known fact, but gods evolve. There is a steady and gradual transformation through human history not only from polytheism to monotheism, but from gods who are touchy, foolish, randy and greedy people, who just happen to be immortal, to disembodied and virtuous spirits living in an entirely different realm and concerned mainly with virtue. Contrast the vengeful and irritable Jehovah of the Old Testament with the loving Christian God of today. Or philandering, jealous Zeus with the disembodied and pure Allah; or vengeful Hera and sweet Mary.

The gods in hunter-gatherer societies manage without priests, and have little in the way of consistent doctrine. The gods of early settled societies, though organised, codified and served by specialised personnel with rituals, were (in the words of Nicolas Baumard and Pascal Boyer) 'construed as unencumbered with moral conscience and uninterested in human morality'. This moral indifference characterised the gods of Sumer, Akkad, Egypt, Greece, Rome, the Aztec, Mayan and Inca empires, and ancient China and India.

Only much later, in certain parts of the world – apparently those places where sufficiently high living standards caused some people to yearn, like hippies, for ascetic purity and higher ideals – did the gods suddenly become concerned with moral pre-scription. Priests discovered that demanding ascetic self-sacrifice induced greater loyalty. Sometimes the switch happened through a reformation, as in Judaism and Hinduism; more often through the emergence of a new and morally prescriptive god cult, as in Jainism, Buddhism, Taoism, Christianity and Islam. These moral gods proved very jealous, and more or less elbowed aside not only the morally neutral religions, but those moral codes that lacked superstitious belief, such as Pythagorism, Confucianism and stoicism. Remarkably, they all seem to recommend some version of the golden rule – do as you would be done by – as illustrated by precepts of Buddhism, Judaism, Jainism, Taoism, Christianity and Islam. They thrived, argue Baumard and Boyer, by appealing to human instincts for reciprocity and fairness – by emphasising proportionality between deeds and supernatural rewards, between sins and penance. In other words, gods evolved by adapting themselves to certain aspects of human nature, the environment in which they found themselves. They were doubly man-made, unconsciously as well as consciously: human-evolved as well as human-invented.

Just as Rome was ripe for Christianity, the same is true of Arabia and Islam. The vast Arab empire was bound to spawn a universal religion of its own, and probably one that was jealous of others, but that it should be Mohamed's version that would win the prize was far from inevitable. (Religion is predictable; religions are not.) Yet in this case, we are assured, it happened the other way round: a religion spawned an empire. In AD 610 Mohamed received the Koran from an angel, while living in a pagan desert town called Mecca, which was a thriving crossroads of the caravan trade, and he went on to win a remarkable battle with divine assistance and conquer Arabia. As is often said, we know a great deal more about the biography of Mohamed than we know of the life of other religious founders.

The evolution of the prophet

Or do we? In fact, every one of those biographical facts is doubtful. Except for one brief Christian reference to a Saracen prophet in the 630s, nothing was written down about the life of Mohamed during his lifetime, the first public mention in the Muslim world coming in 690. The detailed biographies were all written two centuries after he died. And what historians can reconstruct about late antiquity in the Near East tells us that Mecca was not a major centre of trade, indeed it is not mentioned till 741. Clearly, too, the Koran was written down not in a pagan society, but in a thoroughly monotheistic one – it has huge amounts of Christian, Jewish and Zoroastrian lore in it. The Virgin Mary features more frequently in the Koran than in the New Testament; as do a few concepts shared with the long-lost Dead Sea scrolls, which would have been obscure in the 600s, and must have been passed down from older traditions. The Koran is too full of details about Jewish and Christian literature to have been a compilation of notions picked up by a trader, let alone one from a pagan and largely illiterate society.

Indeed, there is nothing to tie the life of the Koran's compiler to the middle of the Arabian Peninsula at all, but lots to tie it to the fringes of Palestine and the Jordan valley: names of tribes, identification of places and mentions of cattle, olives and other creatures and plants not found in the Arabian desert. The story of Lot, Sodom and the pillars of salt is mentioned in the Koran in a way that implies it happened locally – and it almost certainly refers to salt features found close to the Dead Sea. The northern part of Arabia, just outside the bounds of the Roman empire, had long been a fertile breeding ground for exiled Jewish and Christian heresies, each drawing on different traditions and some with an admixture of Zoroastrianism from Persia. It is here, many scholars now suggest, that the Koran is actually set.

The traditional alternative requires, of course, a leap of faith. As the historian Tom Holland, author of *In the Shadow of the Sword*, has put it: 'Mecca, so the biographies of the Prophet teach us, was an inveterately pagan city devoid of any large-scale Jewish or Christian presence, situated in the midst of a vast, untenanted desert. How else, then, are we to account for the sudden appearance there of a fully-fledged monotheism, complete with references to Abraham, Moses and Jesus, if not as a miracle?'

To those who do not accept miracles it seems more likely that the Koran is almost certainly a compilation of old texts, not a new document in the seventh century. It is like a lake into which many streams flowed, a work of art that emerged from centuries of monotheistic fusions and debates, before taking its final form in the hands of a prophet in an expanding empire of newly unified Arabs pushing aside the ancient powers of Rome and Sassanid Persia. It is, in Tom Holland's vivid words, a bloom from the seedbed of antiquity, not a guillotine dropped on the neck of antiquity. It contains bits of Roman imperial propaganda, stories of Christian saints, remnants of Gnostic gospels and parts of ancient Jewish tracts.

Holland goes on to speculate about how the Arab civilisation arose, and minted its new religion as it did so. The (bubonic) plague of Justinian in AD 541 devastated the cities of the (Byzantine) Roman and Persian empires, but left the nomads on the southern fringes of both empires relatively unscathed. Nomads have fewer flea-infested rats in their tents than city dwellers do in their houses, which makes plague much less of a problem. In the wake of the plague, parts of the imperial frontier were depopulated, left undefended and ruined, leaving fertile land for the nomads to expand into. A great war between Constantinople and Persia in the early 600s, in which first Persians then Romans triumphed, further exhausted the hegemonic powers and further emboldened the nomadic tribes on the fringes. The Koran contains hints that it is set against this backdrop of a great war, and includes echoes of the campaigns of Heraclius and his attempt to don the mantle of Alexander.

It is only in retrospect that Mohamed is enshrined as a prophet, the Sunni tradition is crystallised and the Hadiths are written down to give Mohamed a realistic and detailed life. By then the Arabs had established a wide empire with firm but brittle self-confidence, and there was clearly a determination to extinguish any hints of intellectual ancestry of Islam within the infidel faiths of Christianity and Judaism. So the sudden, miraculous, *a nihilo* invention of Islam by Mohamed becomes the story that is told. In fact, what was going on in the 690s was that a newly entrenched Umayyad Amir, Abd al-Malik, set about deliberately cultivating the legend of the prophet, naming him for the first time. 'In the name of God. Muhammad is the messenger of God' was stamped on his coins. He did so deliberately to separate his empire's religion from that of the rival Romans, establish that it was not just a reformed version of Christianity, and 'rub Roman noses in the inferiority of their superstitions', to use Tom Holland's words: 'Out of the flotsam and jetsam of beliefs left scattered by the great flood tide of Arab conquests, something coherent – something manifestly God-stamped – would have to be fashioned: in short, a religion.' Thus, Islam was more the consequence than the cause of Arab conquest.

There is nothing uniquely Muslim about this. It is what Christianity and Judaism did also: construct elaborate backstories to obscure their origins. We can see it most clearly in recent religious innovations, like Mormonism and Scientology. Consider the bald facts of the matter about the Church of Latter-Day Saints: in upstate New York in the 1820s, an impoverished amateur treasure-seeker who had stood trial on a charge of falsely pretending to find lost treasure, named Joseph Smith, claimed he had been directed by an angel to a spot where he dug up gold plates on which was written text in an ancient script and language, which he found he could miraculously translate. The plates had since, he said, been put in a chest, but the angel had told him to show them to nobody. Instead he was to publish a translation. Some years later he dictated 584 pages of this translation, which proved to be written in the style of the King James Bible and to be a chronicle of some early inhabitants of North America who had somehow travelled there by ship from Babylon hundreds of years before Christ, but who none the less believed in Jesus.

Of the two possibilities – that this is true, or that Joseph Smith made it up – one is far more plausible than the other. Yet to me there is nothing, except the grandeur granted by the passage of many centuries, to distinguish the implausibility of Mormonism from that of Christianity, Islam or Judaism. After all, Moses too went up a hill and came down with written instructions from God. All religions look man-made to me.

The cult of cereology

I had an epiphany myself on this matter, every bit as numinous as the ones that happened to Moses, Saul of Tarsus or Joseph Smith – well, almost. It came in the early 1990s, when I got involved in the controversy over the origin of crop circles. When I first read about neatly circular patterns of flattened wheat and barley appearing in English fields, it seemed obvious to me that they were likely to be man-made. That somebody had found a way to trample corn down in a neat circle as a joke after going to the pub seemed vastly more plausible than that aliens or some unknown physical forces had suddenly materialised in Wiltshire and gone about their business undetected at night, and only in fields of grain close to roads, for no discernible reason. At the very least it should be considered as a null hypothesis.

So I did the sensible thing. I went out and made some crop circles myself, to see how easy it was to do. My second attempt was good enough to fool a local farmer into a state of high excitement. With a sister and two brothers-in-law, I entered a night-time crop-circle competition organised by some fans of the supernatural and designed to show how hard they were to 'hoax'. Our results and those of the other teams taking part easily proved the opposite: they were easy to make. Yet the crop-circle craze grew and grew, spawning books, films, guided tours and even an institute of 'cereology', with nobody apparently having the courage or the incentive to insist that they were likely to be man-made. Soon some people were making serious money from the cult through books and lectures. The circles were getting more and more elaborate, and more and more obviously man-made. Yet the explanations now focused on things like ley lines, alien spacecraft, plasma vortices, ball lightning, or quantum fields. Some thought they were messages from Gaia to tell humanity to combat global warming. The entire field was pseudoscience of the most blatant kind, as the slightest brush with its bizarre practitioners easily demonstrated.

Imagine, then, my surprise when I wrote about this, gently mocking the irrationality of not thinking they were man-made, and found myself attacked as an idiot for being closed-minded about supernatural causes. The problem was, you see, that I was ignoring the 'experts' in crop-circle science who said I was wrong. I found I was treated like a heretic: one or two of the attacks were quite vicious. Journalists working not for tabloids but for *Science* magazine, and for a television documentary team, meekly repeated the patently false argument of the self-appointed 'cereologists' that it was highly implausible that crop circles were all man-made – they imbibed the argument from authority with consummate ease. I learned for the first time about the stunning gullibility of the media, and its unthinking reverence for any voice of self-appointed authority. Put an 'ology' after your pseudoscience and you can get journalists to be your tame propagandists. I had watched *Monty Python's Life of Brian*, but had not quite taken in how utterly true to life it was.

A television team did the obvious thing – they got a group of students to make some crop circles one night and then asked a top 'cereologist', Terence Meaden, if they were 'genuine' or 'hoaxed' – i.e. man-made. He assured them categorically on camera that they could not have been made by people. So they told him the circles had been made the night before. The man was pole-axed, and left floundering for words. It made great television. Yet even then, the programme's producer ended the segment by taking the cereologist's side: of course, not ALL crop circles are hoaxes, only this one. Ye gods!

That same summer, two men called Doug Bower and Dave Chorley confessed to having started the whole craze in 1978 after a night at the pub. They gave dates, times, techniques – plenty of convincing detail. A newspaper commissioned Doug and Dave to make one, then asked another top 'cereologist', Pat Delgado, to judge its authenticity. Delgado too made a fool of himself by insisting that the crop circle could not possibly be a 'hoax'. So did the bubble burst? No. The 'cereology' experts promptly went on television to say it was Doug and Dave who were talking nonsense. (Shades of *Life of Brian* again: the true messiah denies he's a messiah.) Everybody just went on believing. In some parts of the country they still do, though I am glad to say the cereologists have gradually faded into obscurity. Even Wikipedia now says crop circles are (mostly) man-made. But the true believers are still out there. A recent book argued that people like me are part of a 'debunking campaign perpetrated by the British Government, the CIA, the Vatican, and their allies in mass media to brainwash the public'.

The temptations of superstition

I have never forgotten that experience – it taught me just how ready people are to believe supernatural explanations, to trust 'experts' (or prophets) even when they are blatantly phony, to prefer any explanation to the mundane and obvious one, and to treat any sceptic as a heretic to be shouted at rather than an agnostic to be persuaded by reason and evidence. Of course, crop circles were too trivial to lead to a whole new religion, but that's my point. Look how easy it is to get a supernatural craze going even with something so banal. In that moment I understood how Joseph Smith and Jesus Christ and Mohamed and many others managed to persuade their followers that they had witnessed divine intervention (whether they had or not). The literary critic George Steiner, in his book *Nostalgia for the Absolute*, argued that people are attracted by higher truths that simplify the world and can explain everything. They are nostalgic for the doctrinal simplicities of medieval religion.

The central theme of the origin of religions is that they are man-made, like crop circles, but also that they have evolved. They are much more spontaneous phenomena than legend later admits. Like technological innovation, they are the result of selection among variants, of trial and error within cultural experiments. And their characteristics are chosen by their times and places. They are also glimpses into just how gullible we are about prescriptive explanations of the world.

It's not just in its theology that religion is a top–down phenomenon, but in its human organisation too. Religions always and everywhere insist upon the argument from authority. You should do this or that because the Pope or the Koran or the local priest says you should. For centuries most of the world convinced itself that the only reason people act morally is because of instruction, that in effect without superstition there can be no ethical behaviour. Priests are continually insisting that there is a link between observance and outcomes, between prayer and good fortune or between sin and illness. In the seventeenth century, generals like Oliver Cromwell attributed their success in battle entirely to divine intervention with just as much insistence as had the heroes of the Trojan War. This was not always a good strategy. The first-century Chinese Emperor Wang Mang fell from power largely because he spent all his efforts trying to follow the portents of heaven, rather than the needs of people.

Nor is skyhook thinking confined to 'god' religions. It animates all sorts of movements that have faith at their heart, from Marxism to spiritualism, from astrology to environmentalism. The reluctance to accept coincidence lies at the heart of telepathy, spiritualism, ghosts and other manifestations of the supernatural. The mystical mentality insists that something caused a coincidence; something made things go bump in the night.

Superstition can be very easily aroused, and not just in people. The psychologist B.F. Skinner kept pigeons in a cage where a machine produced food at regular intervals. He noticed that some of the pigeons seemed to become convinced that whatever they had been doing just before the food appeared was the cause of the food appearing. The pigeons therefore repeated these habits. One bird turned anti-clockwise. Another thrust its head into a corner. A third tossed its head. Skinner felt that the experiment 'might be said to demonstrate a sort of superstition', and reckoned there were many analogies in human behaviour.

Human beings are plainly highly susceptible to superstition. We are ready to attribute agency to inanimate objects at the drop of a hat, and to believe that crystals have healing powers, old buildings house ghosts, certain people are capable of witchcraft, some foods have magical health properties, and somebody is watching over us. It makes evolutionary sense for people to have this intentional stance, because it must have saved lives in the Stone Age. You probably lived longer if you treated every rustle in the grass or every sudden sound as suspicious, and made by a potential enemy. And if occasionally this led you to mistake natural coincidences for malevolent spirits – well then, no harm done. Various so-called neuro-theologists have claimed to find evidence for exactly where this hyperactive intention-detector lies within the brain, or which gene variants make it more hyperactive in some of us than in others. So far there are few consistent results.

But the truth is we all have it to some degree or other, which is why religious belief is found in every part of the world and every age of history, while rational scepticism is a rare and often lonely stance that leaves Lucretius, Spinoza, Voltaire and Dawkins as heretics. Indeed, the paradox of this realisation is that if belief (in the broad sense of the word) is universal, then no amount of argument can extinguish it, and in a sense therefore, gods really do exist – but inside our heads rather than outside. For this reason, neuro-theology is actually rather popular among believers, who take the view that it emphasises the futility of atheism, rather than that it means gods are made up.

Vital delusions

And the inevitable consequence is that, as G.K. Chesterton said, when people stop believing in something, they don't believe in nothing, they believe in anything. It cannot be a coincidence that the decline in Christian worship in Europe has been accompanied by a rise in all sorts of other superstitions and cults, including those of Freud, Marx and Gaia. Indeed, before I turn smug and mock astrology, telepathy, spiritualism and Elvis worship, I should candidly admit that scientists are as prone as any of us to this tendency towards belief. I have become steadily less, rather than more, confident in my ability to distinguish pseudoscience from true science. I am close to certain that astronomy is a science; astrology is a pseudoscience. Evolution is science; creationism is pseudoscience. Molecular biology is science; homeopathy is pseudoscience. Vaccination is science; vaccination scares are pseudoscience. Oxygen is science; phlogiston was pseudoscience. Chemistry is science; alchemy was pseudoscience. I am also pretty sure that the belief that the Earl of Oxford wrote Shakespeare is pseudoscience. So are the beliefs that Elvis is still alive, Princess Diana was killed by MI5, JFK was killed by the CIA, and 9/11 was an inside job. So are ghosts, UFOs, telepathy, the Loch Ness monster, alien abductions and pretty well everything to do with the paranormal.

But more controversially, I also think a lot of what Freud said was pseudoscience. As Karl Popper noted in his essay 'Conjectures and Refutations', the ideas of Marx, Freud and Einstein were all, when he was growing up in Vienna, powerfully explanatory. But he quickly realised that gathering verifications of them was not the way to find out if they were correct. The key was whether they were refutable. Whereas Einstein's ideas could be falsified by a simple experiment, nothing seemed to faze Marxists or Freudians (or the followers of Adler, among whom Popper initially numbered himself). Any event seemed adaptable to fit the theories of Marx or Freud. It was precisely because they always fitted any fact that, in the eyes of their admirers, the theories were so strong: 'It began to dawn on me that this apparent strength was in fact their weakness.' The giveaway was when the theories were refuted by events, and their adherents simply explained away the misfit. In the case of Marx, prediction after prediction about how and where the revolution would happen turned out to be wrong, but followers of Marx repeatedly re-interpreted both the theory and the evidence. 'In this way they rescued the theory from refutation; but they did so at the price of adopting a device which made it irrefutable.'

For me, the characteristic features of a mystical and therefore untrustworthy, theory are that it is not refutable, that it appeals to authority, that it relies heavily on anecdote, that it makes a virtue of consensus (look how many people believe like me!), and that it takes the moral high ground. You will notice that this applies to most religions.

Just like religion, science as an institution is and always has been plagued by the temptations of confirmation bias. With alarming ease it morphs into pseudoscience, even – perhaps especially – in the hands of elite experts, and especially when predicting the future and when there's lavish funding at stake.

One form of superstition that is in headlong retreat is 'vitalism'. This is the old idea that there is something peculiar and special about living tissue. As well as containing carbon and hydrogen and oxygen and all that jazz, a living cell is sup-posed to have some mysterious vital ingredient that makes it 'alive'. Vitalists have been falling back for centuries. The arti-ficial synthesis in 1828 of urea, a substance produced hitherto exclusively by living creatures, was one such blow, which largely destroyed the idea that chemistry was going to find the vital prin-ciple. Vitalists fell back on physics, and later quantum physics, where they suggested that mysterious peculiarities might still exist. But that too was blown away by the discovery of the struc-ture of DNA. In a way you could argue that the double helix did confirm that there is something peculiar and special about living tissue – namely, that it contains digital information cap-able of both replicating itself and instructing the synthesis of machinery for harnessing energy. The secret of life, unexpectedly, turned out to be an infinitely combinatorial message written in digital form in three-letter words in a four-letter alphabet. This was very much not what vitalists had expected; it seemed too mundane – though actually it is one of the most beautiful ideas ever to cross a human mind – that life is information. And so, with the elucidation of the genetic code in 1966, Francis Crick confidently declared vitalism dead and buried.

Only it still lives on in various pseudosciences. Homeopathy is based on vitalism. Its founder Samuel Hahnemann believed that diseases 'are solely spirit-like (dynamic) derangements of the spirit-like power (the vital principle) that animates the human body'. Organic farming also originated in vitalism, its founder Rudolf Steiner believing that in order 'to influence organic life on earth through cosmic and terrestrial forces', it was necessary to 'stimulate vitalizing and harmonizing processes in the soil', an insight he acquired through clairvoyance. The preparations necessary to achieve this consisted of various materials placed inside cow horns and buried in ritual fashion so that they could act as antennae to pick up cosmic vibrations. These 'biodynamic' superstitions have largely faded from the mainstream organic movement, but its faith in certain agricultural technologies, like copper sulphate pesticides, but not in others, like genetic modification, remains essentially mystical.

The climate god

The theory that industrial emissions of carbon dioxide will in the future cause dangerous global warming, though far more scientific than these superstitions, has also acquired overtones of religiosity, as anybody who questions it quickly finds. That carbon dioxide is a greenhouse gas is not in doubt, and other things being equal, increases in carbon dioxide levels will cause warming. Such warming, so the theory goes, is not dangerous in itself, but will be greatly amplified by extra water vapour released into the atmosphere by the initial warming, and may then be sufficiently large and rapid as to threaten global catastrophe. It will also overwhelm any natural changes in climate that occur.

In that sense, carbon dioxide emissions are the 'control knob' of climate.

This is a huge subject and beyond the scope of this book, but an increasing number of scientists tell me they are worried that this is too top–down a perspective – that carbon dioxide levels are just one influence among many, including 'internal variability' that has no external cause. This explains, these sceptics (such as Judith Curry of the Georgia Institute of Technology) think, the failure of the climate to warm nearly as fast over recent decades as predicted. It also explains the fact that Antarctic ice cores reveal a clear relationship between temperature and carbon dioxide as the earth goes into and out of ice ages that is the reverse of that predicted by the theory: carbon dioxide levels follow temperature up and down, rather than precede them. Effects cannot precede causes, and we now know almost for sure that ice ages are caused by changes in the earth's orbit, with carbon dioxide playing a minor, reinforcing role, if any at all. In short, there is a tendency to over-prioritise carbon dioxide as a cause of global temperature, rather than just another influence among many.

Simplistic cause-seeking is characteristically religious. Certainly, when doubters make the arguments above they are often met with a series of largely religious arguments: that they are 'deniers' of the truth, that their position is morally wrong because it ignores the needs of posterity, or that they should accept the majority consensus. But the whole point of science, the whole thrust of the Enlightenment, is the rejection of arguments from authority. Science, said Richard Feynman, is the belief in the ignorance of experts. Observation and experiment trump scripture. To hear even some scientists, at least in the field of climate, insisting that there is only one true voice of authority, is to be reminded of religion, not enlightenment. Besides, there is near scientific unanimity only that there will be some warming, not that it will be dangerous.

Another religious argument that comes up is that, yes, catastrophic warming may be unlikely, but if there is even a minuscule chance of it, then almost anything we can do, however painful, to forestall it will be worthwhile. This is a form of Pascal's wager: Blaise Pascal argued that even if God is very unlikely to exist, you had better go to church just in case, because if he does exist the gain will be infinite, and if he does not the pain will have been finite. To me this is a dangerous doctrine, which justifies inflicting real pain in the here and now on disadvantaged people on the basis of forestalling a distant possibility of doom. This was exactly the argument used by eugenicists: the noble end justifies cruel means. Besides, Pascal's wager applies to every other possible disaster, and it applies just as much to the means as to the ends. What if renewable energy rolled out on a grand scale proves so environmentally damaging that it does great harm? Bio-energy, a policy intended to forestall global warming, is already killing hundreds of thousands of people each year by putting up the price of food.

Various dissenting sceptics, from the late Michael Crichton to the Nobel-winning physicist Ivar Giaever, from former Australian Prime Minister John Howard to former British Chancellor Nigel Lawson, have also drawn analogies with just how religious climate-change arguments are becoming. We are told that we are sinning (by emitting CO_2), that we have original sin (human greed), which has banished us from Eden (the pre-industrial world), for which we must confess (by condemning irresponsible consumerism), atone (by paying carbon taxes), repent (insisting that politicians pay lip service to climate-change alarm), and seek salvation (sustainability). The wealthy can buy indulgences (carbon offsets) so as to keep flying their private jets, but none must depart from faith (in carbon dioxide) as set out in scripture (the reports of the Intergovernmental Panel on Climate Change). It is the duty of all to condemn heretics (the 'deniers'), venerate saints (Al Gore), heed the prophets (of the IPCC). If we do not, then surely Judgement Day will find us out (with irreversible tipping points), when we will feel the fires of hell (future heatwaves) and experience divine wrath (worsening storms). Fortunately, God has sent us a sign of the sacrifice we must make – I have sometimes been struck by the way a wind farm looks like Golgotha.

When Rajendra Pachauri resigned as chairman of the supposedly neutral and scientific IPCC in February 2015, his resignation letter to the UN Secretary General included the remarkable admission: 'For me the protection of Planet Earth, the survival of all species and sustainability of our ecosystem is more than a mission. It is my religion and my *dharma*.' In the words of the left-wing French philosopher Pascal Bruckner, who is highly critical of climate policy, 'The environment is the new secular religion that is rising, in Europe especially, from the ruins of a disbelieving world.' He writes that 'The future becomes again, as it had once been in Christianity and communism, the great category of blackmail.'

I am being a little tongue-in-cheek here. I do not really think that climate-change enthusiasts consider Al Gore to have divine properties. And yes, there is real scientific evidence to support some possibility of alarm. But I am pointing out that there is a long-standing human tradition to become so enthusiastic about a favoured scientific, religious or superstitious explanation for the world as to close your mind and come to hate those who disagree. We have seen it far too often to ignore it, and scientists have shown themselves no better than the rest of us at resisting the temptation.

The weather gods

When southern England experienced widespread floods in the winter of 2013–14, a local politician in the UK Independence Party named David Silvester was heard to muse that it must be God's punishment for the country enacting a law that allowed gay marriage. He was rightly mocked. But a matter of days later, pretty well every normal politician, with a few exceptions, was blaming the floods on man-made climate change, even though there had been no net warming for fifteen years, and there was no evidence of significant trends in extreme weather or the wetness of British winters, and there was plentiful evidence for changes in land use and dredging policy as the cause of the flooding. Indeed, a study by Southampton University scientists concluded that any increase in flooding in Britain was caused by urban expansion and population growth, rather than climate change. The Met Office agreed that 'there continues to be little evidence that the recent increase in storminess over the UK is related to man-made climate change'.

In the face of this scientific brick wall, activists tended to fall back on vague phrases like 'consistent with'. Floods may not be directly attributable to climate change, but the pattern is consistent with it. This is the language of religion. As Nigel Lawson puts it:

> So what? It is also consistent with the theory that it is a punishment from the Almighty for our sins (the prevailing explanation of extreme weather events throughout most of human history). Indeed, it would be helpful if the climate scientists would tell us what weather pattern would *not* be consistent with the current climate orthodoxy. If they cannot do so, then we would do well to recall the important insight of Karl Popper – that any theory that is incapable of falsification cannot be considered scientific.

So when every storm and flood of recent years, every typhoon, hurricane and tornado, every drought and heatwave, every blizzard and ice storm is attributed (mostly by politicians rather than scientists) to man-made climate change, ignoring all the other factors that have contributed – including man-made ones such as vegetation changes or changes to land drainage and development – what is the difference from the man who blamed it on gay marriage? Both are attempts to turn the weather into the wages of sin.

The human tendency to seek intentional explanations of the weather is as old as time. 'Each natural event is supposed to be governed by some intelligent agent,' wrote David Hume. Somewhere deep in our psyches, we have just never really accepted that a thunderstorm does not have an agency behind it, that a drought is not a punishment for some misdemeanour. It's the intentional stance again. In the old days it was Zeus or Jehovah or the rain gods. In the sixteenth century it was witches: the historians Wolfgang Behringer and Christian Pfister discovered that organised witch-hunting and the burning of supposed witches as scapegoats in Europe correlated neatly with episodes of bad weather and failed harvests during the climate cooling known as the Little Ice Age. Peasant communities suffering damage from climatic change often pressed authorities for the organisation of witch-hunts.

Even in the eighteenth century any natural disaster was assumed by most people and most leaders to be divine retribution for sin: Leibniz's theodicy demanded it. For a brief lull in the twentieth century the rational view prevailed that weather was just weather, and nobody's fault. But with the new tendency to blame every storm and flood on emissions of carbon dioxide, that lull is over, and the sigh of relief that we can go back to blaming each other for the weather is almost audible. The huge appeal of the 'extreme weather' meme of recent years comes from the fact that it plays into this divine-retribution mentality.

The most important fact about extreme weather is that the number of deaths caused by floods, droughts and storms has dropped by 93 per cent since the 1920s, despite a trebling of the world population: not because the weather has grown less wild, but because the world has grown rich enough to enable us to protect ourselves better.

15

The Evolution of Money

And trickles of silver and gold, also copper and lead, would stream
And pool in the earth's hollows. When cooled, men saw the gleam
Of their glinting colours in the soil, and drawn to what they'd found –
The shiny smoothness of the nuggets – pried them from the ground,
And saw these bore the shapes of the depressions where they lay.
Then this drove home that they could shape the nuggets in this way –
Melting them down and pouring them into any mould they made.

Lucretius, *De Rerum Natura*, Book 5, lines 1255–61

Money is an evolutionary phenomenon. It emerged gradually among traders, rather than being created by rulers – despite the heads of kings on the coins: those just illustrated the tendency of the powerful to insist on monopolies. And there is absolutely no reason why money must be a government monopoly. There's a story that illustrates this, from the dawn of Britain's Industrial Revolution. In the eighteenth century more and more poor people started moving to towns and working for wages rather than staying in their rural villages and being paid in kind by their semi-feudal employers. This presented employers with a new problem – a shortage of coins. There were gold guineas in circulation for the rich to use, but too few silver crowns or shillings, or copper pennies or halfpennies. Silver coins were worth more, in gold, in China than at home, so they tended to be melted down and shipped east, while the Royal Mint sniffily refused to mint more for most of the eighteenth century. Existing silver shillings were deteriorating in quality. As for the Bank of England, it would issue no paper notes smaller than £5. The entrepreneurs of Birmingham, unable to pay wages in silver, found too few copper pennies available and resorted to using counterfeits, which were abundantly if illegally supplied to them in the back streets.

One Birmingham businessman, Matthew Boulton, the owner of the giant Soho works, petitioned Parliament to let him solve the problem by granting him the right to produce new regal coins, but the Royal Mint was as jealous of its monopoly as it was complacent about the problem, and Boulton was rebuffed. Another businessman in Wales, Thomas Williams, had a better idea. After striking coins with lettered edges that would be hard to clip, he tried to interest the Royal Mint in the new designs. No response. So in 1787 he began producing copper coins from his mine at Parys in Anglesey. He did not pretend they were pennies, but merely 'tokens' that could be exchanged for pennies, which was legal. The copper tokens were called 'druids'. Beautifully designed and smoothly executed, they had a low relief of a hooded, bearded druid on one side, wreathed in oak leaves, with the letters 'PMC' – for Parys Mine Company – on the other side, and around the edge the legend: 'WE PROMISE TO PAY THE BEARER ONE PENNY'. What made the coins especially hard to fake or clip was the writing on the outside of the raised rim: 'On demand, in London, Liverpool, or Anglesea'. The owners of factories started paying their workers in druids, and local shopkeepers started accepting them in lieu of pennies. It was an entirely private currency.

John Wilkinson, an ironmaster in Staffordshire with a large and growing business, then asked Williams to strike coins for him to pay his workforce. These coins were known as Willeys, after the Wilkinson ironworks at New Willey. But Wilkinson's coins were half the weight of Williams's, so his workers soon found that tradesmen accepted them as halfpennies, not pennies. That they carried an image of the ironmaster Wilkinson's own profile caused mockery in London:

As Iron when 'tis brought in taction,
Collects the coppers by attraction,
So thus, in him, was very proper
To stamp his brazen face on copper.

Other entrepreneurs followed suit. Soon (in a reversal of Gresham's Law, that bad money drives out good) the tokens had driven out the counterfeit coins and become a legitimate currency, preferred to the sovereign coins and accepted even in distant London. The habit of striking private coins was catching on. In 1794, sixty-four tradesmen issued coins for the first time. By 1797 over six hundred tons of tokens were in circulation. Private coiners had solved the problem of a shortage of change. In effect, as George Selgin – the outstanding historian of this curious episode in his book *Good Money* – puts it, Birmingham businessmen had privatised the penny. Their coins were a vast improvement on the Royal Mint's rivals. This despite the fact that the new coins had been designed from scratch in just a few years, and had no legal protection against fraud, unlike the Mint's coins. Unprotected by monopoly privilege, the commercial coiners had not only to be cost-effective, but to attract the best engravers and strikers, and had to design their coins so they would be hard to imitate. 'Such concerns,' says Selgin, 'were utterly foreign to the denizens of that cluttered cloister that was the old Tower Mint.'

The Mint had not only refused to produce enough coins to service the new industrial economy, it had refused to adopt modern methods. As Selgin observes, 'Nothing better illustrates the tenacity with which the Company of Moneyers resisted technical innovation than their successful scuttling, over the course of more than a century, of repeated attempts to mechanize coinage through the substitution of screw or roller presses for shears and hammers.'

The private coiners now took a step too far. In 1797 Matthew Boulton had at last won the right to strike regal copper pennies with steam-powered presses – to a design with a raised rim that gave them their nickname of 'cartwheels'. But when in 1804 he began striking silver coins (or rather re-striking silver Spanish dollars as English five-shilling coins), the slumbering Royal Mint eventually awoke, and stirred Parliament into action to defend its monopoly. It adopted Boulton's methods, lobbied to win back coinage contracts, and gradually regained its monopoly. So it was that an ancient and hidebound institution was modernised, not by direction, but by competition.

Private token coins had one last hurrah in 1809–10, when bad harvests – necessitating imports of grain from the Continent through Napoleon's blockade, paid for with gold and silver – and the costly demands of the Peninsular War caused an acute shortage of silver coins in the British Isles. Once again, metal entrepreneurs and, this time, bankers began striking silver shilling and sixpence tokens as well as copper pennies. This time the politicians, with their usual preference for crony monopoly, objected, and by 1814 private token coins were banned by law. The result was a predictable shortage of coins, because the Mint was not ready for several years to produce enough regal coins. To fill the vacuum, counterfeit coins and French coins began circulating again. An employer wishing to pay wages in 1816 had to make do with a mixture of old bank tokens, Boulton coppers, maybe a few worn druids and Willeys, some French sous or Spanish dollars, or counterfeit coins. Selgin concludes: 'Such were the alternatives to commercial money for which Parliament, in its blind impetuosity, had cleared the way.'

The Scottish experiment

There is an even more persuasive example of monetary evolution from north of the Scottish border. Between 1716 and 1844

Scotland experienced unparalleled monetary stability, pioneering financial innovation and rapid economic growth as it caught up with England. It had a self-regulating monetary system, which worked as well as any other monetary system at any other time or place. Indeed, it was so popular that Scots rushed to praise and defend their banks – a phenomenon largely unheard of in history.

Under the Act of Union in 1707, the Scots dropped their currency – the 'pound Scots' – in favour of the English pound. At first there continued to be a monopoly central bank with the power to issue currency: the Bank of Scotland, founded in 1695, the year after the Bank of England. But later, Parliament in London, concerned at Jacobite influence in the Bank of Scotland after the Old Pretender's revolt of 1715, gave its rival, a private institution called the Royal Bank, the right to issue currency. Initially there was war between the two banks – each hoarding the other's notes, then presenting them in large amounts to trouble the issuer. Peace then broke out, and the two rival banks eventually agreed to accept each other's notes and exchange them regularly. They were later joined by other note-issuing banks, including the Clydesdale, the Union Bank of Scotland, the North of Scotland Bank, the Commercial Bank of Scotland, the British Linen Bank and many more. In other words, the value of a particular piece of paper money depended on the fragile reputation of one of these private companies, none of which had monopoly power. Surely this was a recipe for disaster?

Quite the reverse. Each of the issuing banks remained keen to have its rivals accept its notes, so took a cautious and sensible approach to lending. The notes were exchanged twice a week, so any doubts about bad lending decisions would be quickly revealed if the exchange system broke down. The system was self-regulating through competition. Banknotes grew more popular, not less, and soon Scotsmen preferred them to gold guineas: more convenient and just as trustworthy. The country came to depend on paper money more than any other. The Scottish banking system proved efficient, innovative, stable and calm. It required only slim precious-metal margins of 1–2 per cent, and introduced numerous new features such as the cash-credit account, branch banking and interest on small deposits. Unlike in England, the banks issued convenient notes of one pound or even less – and some even accepted pound notes torn in half as being worth ten shillings (i.e. half a pound).

Scottish banks sailed comfortably through the crisis of the 1745 Young Pretender's rebellion, when the rest of Scottish society was torn asunder, without financial discomfort. For over a century the system thrived. There were half as many bank failures in Scotland as in England, and they all paid their losses in full. Just £32,000 was lost in bank failures during this period, whereas in England as much as that was sometimes lost in a single year. One high-profile failure, of the aptly named Ayr Bank in 1772, showed how the system of self-regulation worked. The Ayr Bank's aggressive lending was distrusted by its rivals, so they had avoided entangling themselves with it. Instead the Ayr Bank borrowed from London banks, including the Bank of England. It went bust because of a series of bank runs starting in London, which took down more than twenty prominent banking houses. Because it had been avoided by the main Scottish banks, the Ayr Bank's failure took only a few local Scottish banks with it. The main issuing banks acted as lenders of last resort to smaller banks during the crisis, which not only saved them but gave the whole system future credibility. Even the Ayr Bank paid off its creditors eventually, with the massive sum of £663,397.*

---

* The 1772 financial crisis led indirectly to the American Revolution, both because it drew large amounts of gold out of America to repay debts in London and because it caused the East India Company to default on a Bank of England loan; to recover its position, the company sought to sell its warehoused tea, which it dumped in the colonies, with help from the government's Tea Act of 1773 to enforce the company's monopoly on tea sales. This led to the Boston Tea Party. American liberty, and the great thinking that led to the Constitution, got its opportunity bottom–up from a financial and commercial crisis, in other words.

Malachi Malagrowther to the rescue

As the 1772 financial crisis demonstrated, England during this period was plagued by frequent bank failures and credit crises, despite having a monopoly issuer of currency, the Bank of England, with lender-of-last-resort responsibilities. Yet, instead of looking across the border with the intention of emulation, politicians kept trying to make the Scottish system more like the English one. In 1765 the Scottish banks were banned from issuing small notes worth less than £1, despite there being no evidence that they caused any trouble. In 1826, following yet another severe English banking crisis, although no Scottish bank had failed, the Chancellor of the Exchequer Robert Peel tried to ban the issuance of Scottish notes worth less than £5. He (or rather the jealous Bank of England) was disturbed that these notes were circulating in parts of northern England.

Peel was seen off by an unlikely opponent. The great Scottish poet and novelist Sir Walter Scott, writing under the pseudonym 'Malachi Malagrowther', inveighed against Peel's attempt at nationalisation of Scotland's monetary system. He called it 'this violent experiment on our circulation – demanded by no party in Scotland – nay, forced upon us against the consent of all who can render a reason, fraught with such deep ruin if it miscarry, and holding forth no prospect whatever of good even should it prove successful'. Since the Act of Union only allowed a change for the 'utility to the subjects of Scotland', Peel was forced to appoint two parliamentary inquiries, which found nothing wrong with the Scottish banking system. It was 'a system admirably calculated to economise the use of Capital to excite and cherish a spirit of useful Enterprise, and even to promote the moral habits of the people, by the direct inducements which it holds out to the maintenance of a character for industry, integrity and prudence'.

In 1844 Peel, by now Prime Minister, tried again, and this time he managed effectively to buy the support of the chief Scottish banks by offering them a comfortable cartel in exchange for regulation by the Bank of England. The consequence was almost immediate. Under the morally hazardous umbrella of a central bank, irresponsible banking appeared in Scotland. By 1847, Scotland's banks were indeed 'fraught with ruin' because of bad lending and needed bailing out by the Bank of England. Peel's act had indeed 'miscarried', and was suspended. Malagrowther was absolutely right.

Financial stability without central banks

If Scotland is not to your taste, try Sweden. In the nineteenth century Sweden had a free banking system, in which banks competed to issue their own paper currencies. The effect of this system: 'During the seventy years of its existence, not a single bill-issuing bank failed, no bill-owner lost a krona, and no bank had to shut its windows for even a single day,' as recounted by Johan Norberg, citing Per Hortlund.

Or Canada in the 1930s. Which advanced economy survived the Great Depression in the best shape and had the least trouble in its banking system? The one with no central bank: Canada.

Or indeed the United States. American state banks issued currency throughout the nineteenth century, but during the Civil War the federal government tried to raise funds by allowing federally chartered banks, so long as they backed their issuance with government bonds. With disappointingly few takers, the government hit the state banks with a 10 per cent tax on outstanding banknotes, effectively killing their role. When the government paid down its debt in the 1880s, the bond-security provision caused the number of notes in issue from national banks to fall. The obvious answer, to free the banks to issue notes as required based on their assets and let the market regulate them, as happened in Canada, was blocked by William Jennings Bryan, the populist Democrat. He frustrated every attempt to free the national banks, and President Grover Cleveland's attempts to repeal the 10 per cent tax on state banks. Bryan continued his crusade against the asset currency into the first decades of the twentieth century, and eventually reformers turned instead to the idea of a central bank with exclusive powers to issue notes. So Bryan's long resistance to what he called the monopoly of the banks led directly to the creation in 1913 of a true monopoly of one bank, the Federal Reserve. Nassim Taleb points out that when Ron Paul, as the libertarian presidential candidate, called for the abolition of the Fed, he was called a crank; but had he called for the setting up of a monopoly with the power to price any other commodity than money, he would have been called a crank for that.

In short, there is no question that a country can run a stable paper currency without a gold standard, a central bank, a lender of last resort, or much regulation; and not only avoid disaster, but perform well. Bottom–up monetary systems – known as free banking – have a far better track record than top–down ones. Walter Bagehot, the great nineteenth-century theorist of central banking, admitted as much. In his influential book *Lombard Street*, he effectively conceded that the only reason a central bank needed to be a lender of last resort was because of the instability introduced by the existence of a central bank.

The history of central banking bears this out. The Bank of England was created in 1694. By 1720 Britain was in its most desperate financial crisis, the South Sea Bubble, a speculative fraud based around persuading people to swap government debt for shares in a trading company that never traded. Rather than take away the punchbowl as the party got going, the Bank of England enthusiastically tried to join in, by submitting a rival bid to take over the national debt and issue shares.

From 1718, John Blunt, the South Sea Company's main mover, had himself modelled its strategy – which consisted of little more than ramping up the share price and living high on the hog on investors' money – on a similar French scheme. The monopoly French government bank, the Banque Royale, created by the Scottish murderer, gambler and brilliant entrepreneur John Law, became the national bank of France. The Regent, Philippe, duc d'Orléans, granted Law sweeping economic powers, which he used to suck as many rich people as possible into a bubble in shares in the Mississippi Company, which his bank owned and which had a monopoly in trade with North America and the West Indies. By talking up the riches of Louisiana, Law created a bubble in the bank's shares, which eventually burst.

The contrast with what happened in Law's native Scotland is acute. Both countries introduced paper currencies. The one that granted a bank a centralised monopoly backed by the state ruined everybody. The one that plumped for a decentralised, evolutionary system of competition worked beautifully. Central banks tend to behave pro-cyclically, pushing down the cost of borrowing as credit expands and slamming the door shut when it contracts – just as they did in the early 2000s. By contrast, decentralised money has a far better record.

There is even a fascinating parallel with the Scottish free-banking experiment going on today. Three countries – Panama, Ecuador and El Salvador – have 'dollarised' their economies, by deciding to use the dollar as their currency. This means, of course, that their banks are without a lender of last resort, because the US Fed is not likely to bail out a Panamanian bank. The consequence of this has been surprisingly positive. With moral hazard gone, the banks in the three dollarised countries have behaved cautiously, so much so that Panama's banks are now considered highly stable, and the International Monetary Fund has stated that the very lack of a lender of last resort has 'contributed to the resilience and stability of the system'. The IMF sees the need for some kind of permanent liquidity facility, but instead of a central bank it proposes for El Salvador a pool drawn from all banks' current reserve requirements, with a penal rate of interest for those banks that need to use the facility. This is not unlike the system that Scotland operated with such success for so long.

The China price

Surely, though, the great financial crisis that began in 2008 was caused by too little regulation, and too much greed? So at least goes the conventional wisdom. The repeal of the Glass-Steagall Act (which separated banking and securities trading) in 1999 was the culmination of a decade of financial deregulation, according to this view. Like so much conventional wisdom, this is almost wholly wrong.

As the author George Gilder comments, in the run-up to the crisis, 'every large institution was thronged with examiners, overseers, supervisors, inspectors, monitors, compliance officers and a menagerie of other regulatory constabulary'. These invariably gave the institutions a clean bill of health right up till the moment they declared them in need of bail-out. The Independent National Mortgage Corporation, which collapsed in 2008, costing $11 billion to the FDIC plus losses to depositors and creditors, had hosted up to forty government examiners on site, all of whom gave Indymac high ratings. AIG, whose credit default swaps almost killed the world economy the same year, had been, in Gilder's words, 'supervised and pettifogged by federal, state, local, and global beadles galore, in fifty states and more than a hundred countries'. My own experience as chairman of a bank was of endless reassurance from intrusive and detailed regulation right up till the point when it all went wrong. Far from warning of the crisis to come, regulators did the very opposite, and gave false reassurance or emphasised the wrong risks.

Indeed, the problem is worse than that. The crisis of 2008 was triggered to a large extent by a top–down interference in something that should have been a bottom–up system: credit. Greed, incompetence, fraud and error were in abundant supply, but they always are. A plethora of regulations encouraged and rewarded them.

Consider the ingredients of the crisis. Like almost all financial crises in history, the immediate cause was the bursting of an excessive bubble in asset prices, especially in property prices. This was true of East Asia in 1997, Japan in 1989 and various crises in the 1970s, 1920s and earlier decades and centuries. The key to understanding the 2008 crisis is how the bubble was inflated.

First, the Chinese government, by radically devaluing its currency in 1994 to stimulate a mercantilist export strategy, and holding it down thereafter to keep exports competitive, created huge global imbalances between Eastern savers and Western borrowers. In effect, the Chinese made their exports competitive and invested the proceeds in cheap loans to Westerners. Had exchange rates been allowed to find their own levels, currencies and interest rates would have adjusted more smoothly, and Western borrowers would have had trouble funding their mortgage habit cheaply. This is not to pick on the Chinese, but to remind you that politicians, not markets, took a key decision. As the former Congressman and budget director David Stockman put it: 'In a world of massive US current account deficits and the "China price", the American economy was, in effect, importing gale force wage and product deflation. And it would continue to do so until China's rice paddies were drained of excess labor and the People's Printing Press stopped pegging its exchange rate.' By the time the system imploded in 2008, the Chinese central bank owned an astonishing $1 trillion of American residential mortgages.

Second, the flood of cheap debt that flowed through Western economies was bound to find its outlet in asset price inflation, and it did. For nearly four hundred years bubbles have happened when borrowing is cheap, and they will go on happening. At first in the late 1990s it was dot-com stocks that bubbled and popped, then house prices. And as so often, authority, far from discouraging the bubble, actively inflated it. The US Federal Reserve's policy of driving down interest rates to keep the stock market afloat and save Wall Street (rather than Main Street) after the dot-com bust was the single biggest cause of the housing bubble that followed. The Greenspan Put, they called it.

But third, and crucially, there was active, official encouragement of irresponsible lending. American politicians not only allowed banks to lend this cheap money to people with no deposits and little or no capacity to repay; they not only encouraged it; they actively mandated it by law.

How much was Fannie's fault?

The seeds were sown in 1938, when the Roosevelt administration founded the Federal National Mortgage Association, better known as 'Fannie Mae', as a government programme to provide mortgages to moderate-income people whom the banks would not touch. The purpose was to stimulate house-building, even though the housing market had already recovered by the time Fannie Mae got going. It operated by buying home loans from banks for cash, thus running the risk of inducing the banks to offer loans whose creditworthiness they did not worry about. And because Fannie Mae had the credit of the US government behind it, default was no skin off its own nose either. In effect Fannie Mae simply took a fee for giving a government guarantee to a loan, at taxpayer's expense – nice work if you can get it.

In the 1960s President Lyndon Johnson half-'privatised' Fannie Mae as a 'government-sponsored enterprise' (GSE). It was joined in 1970 by its junior brother the Federal Home Loan Mortgage Corporation, known as 'Freddie Mac', but both were left with an implicit government guarantee, which kept their borrowing costs down. This took the form of a line of credit to the Treasury, which everybody knew to be potentially unlimited if necessary. That is to say, the markets assumed that if Fannie or Freddie got into trouble, the taxpayer would bail them out (as indeed she did). In effect the upside was now private, the downside public. Says David Stockman, 'The GSEs were actually dangerous and unstable freaks of economic nature, hiding behind the deceptive and good-housekeeping seal afforded by their New Deal-sanctioned mission to support middle-class housing.'

When Stockman was Ronald Reagan's head of the Office of Management and Budget, he set out to strangle Fannie and Freddie by gradually forcing them to borrow at market rates. Horrified lenders, brokers, builders and suppliers joined in a 'mighty coalition to keep private enterprise humming on cheap, socialized credit'. They lobbied Congress to stop him, and led by the Republicans, it did so. This was a paradigmatic case of crony capitalism in action against the free market.

Meanwhile, commercial lenders were coming under pressure from groups like the Association of Community Organizers for Reform Now (ACORN) to lower their lending standards. ACORN discovered that in the period running up to the completion of a merger, when a date for completion had been set, such banks were highly vulnerable to lawsuits that claimed they were not complying with the 1977 Community Reinvestment Act, which forbade racial discrimination in lending. Refusing loan applicants for lack of a down-payment, or for a poor credit history, tended to hit African-Americans harder than white people. Against a merger deadline such banks would make concessions when sued by ACORN, relaxing their lending criteria, and making grants to ACORN itself to pursue the project of increasing lending to low-income customers – often with ACORN-originated mortgages. Eventually, it was not just small banks that felt this pressure. When Chase Manhattan merged with J.P. Morgan in 2000, both banks donated hundreds of thousands of dollars to ACORN.

However, at this stage Fannie Mae and Freddie Mac still refused to take bad loans on to their books. ACORN went to work to lobby Congress to change the GSEs' mandates. In 1992, under the first Bush administration, they succeeded, and Congress imposed new affordable-housing goals on Fannie Mae and Freddie Mac, requiring them to accept loans with down-payments of 5 per cent or less, and to accept customers with poor credit histories of less than a year. Fannie and Freddie were to use their privileged ability to borrow in the capital markets because of the implicit government guarantee. ACORN drafted the key part of the legislation for the House banking committee chairman.

The Clinton administration effectively made the mandates into a quota system, insisting that 30 per cent of all loans bought by Fannie Mae and Freddie Mac must be to low- and moderate-income borrowers. But thus far the quotas only affected one-quarter of the lending industry. In July 1994 ACORN met President Clinton and persuaded him to extend the low-income lending mandates to non-banks, by insisting that lending criteria not discriminate on the basis of race, even as an accidental by-product of discriminating on the basis of credit risk. Clinton announced the new policy in June 1995, with ACORN as guests at the ceremony.

In 1999 the administration raised the low-income quota to 50 per cent, and the share to those on very low incomes to 20 per cent, and began to get serious about enforcing these targets. It also started offering subsidies to reduce down-payments as part of its 'national home-ownership strategy', a sure way of driving up house prices. Fannie and Freddie, reported the *New York Times*, were 'under increasing pressure from the Clinton administration to expand mortgage loans among low and moderate income people'. This was done specifically so as to 'increase the number of minority and low income home owners' with mortgages.

In short, the explosion in sub-prime lending was a thoroughly top–down, political project, mandated by Congress, implemented by government-sponsored enterprises, enforced by the law, encouraged by the president and monitored by pressure groups. Remember this when you hear people blame the free market for the excesses of the sub-prime bubble. It is simply a myth that the problem came from deregulation. There was a progressive and enormous increase in regulation during the period in question. The (second) Bush administration, for example, added regulations to the US economy at the rate of 78,000 pages a year. It increased the cost of financial regulation by 29 per cent.

From 2000, Fannie and Freddie's appetite for sub-prime loans increased markedly every year, encouraging a rich harvest of increasingly crazy loans by mortgage originators to supply this appetite. House-builders, lenders, mortgage brokers, Wall Street underwriters, legal firms, housing charities and pressure groups like ACORN all benefited. Taxpayers did not. By the early 2000s, Fannie and Freddie were well intertwined with politicians, donating rich campaign contributions especially to Congressional Democrats, and giving rewarding jobs to politicians – Clinton's former Budget Director Franklin Raines would pocket $100 million from his brief spell in charge of Fannie. Between 1998 and 2008, Fannie and Freddie spent $175 million lobbying Congress.

In 2002 Fannie commissioned a report from three economists, Joseph Stiglitz and Peter and Jonathan Orszag, which concluded that the risk to the government from a potential default by Fannie or Freddie because of sub-prime lending was 'effectively zero' – 'so small that it is difficult to detect'. Congressman Barney Frank, in a 2003 speech, said that the two enterprises were 'not facing any kind of financial crisis . . . the more people exaggerate these problems, the more pressure there is on these companies, the less we will see in terms of affordable housing'. The economist Paul Krugman was still insisting as late as July 2008 that 'Fannie and Freddie had nothing to do with the explosion of high-risk lending a few years ago,' and also had nothing to do with sub-prime loans. By contrast, Congressman Ron Paul was already warning that the special privileges granted to the two GSEs meant that 'the losses will be greater than they would otherwise have been had the government not actively encouraged overinvestment in housing'.

Yet the policy continued. By 2008, when it all went wrong, the second Bush administration had raised the quota for low-income loans to 56 per cent. Fannie and Freddie could not find enough good loans to meet the quota even before this, so they had relaxed their underwriting criteria and started accepting more and more sub-prime loans. This exposure to sub-prime had remained concealed from the market, because none of these loans was called sub-prime: the GSE term for them was 'Alt-A' loans, but the difference is purely semantic. So the failure to report this vast lake of sub-prime loans itself contributed to the worsening of the crisis. I remember all too well the attitude of most people in the market at the time: 'Sure, there's some disgracefully irresponsible lending out there, but it's only a small part of the market.' If only. Fannie and Freddie not only funded much of the mis-investment in residential real estate, writes the banker John Allison in his book *The Financial Crisis and the Free Market Cure*, they also 'provided materially misleading information that contributed to errors by other market participants'.

In 2005–07, fully 40 per cent of loans bought by Fannie and Freddie were sub-prime or Alt-A. While house prices were increasing all seemed rosy, especially when new home owners found there was no interest payable for several years, and especially when the price rises allowed defaults to be turned into extra borrowing through refinancing. But eventually defaults began to snowball.

The full extent of GSEs' sub-prime loans only emerged after they went bankrupt and were put into a Treasury Department conservatorship in 2008. By the time they became insolvent that year (shortly after Paul Krugman had said they were not in trouble, worries about them were overblown and they had no sub-prime loans), Fannie and Freddie were holding more than two-thirds of all sub-prime loans, or $2 trillion worth. Nearly three-quarters of new loans passed through their hands that year.

I have dwelt on the story of Fannie, Freddie and the Clinton and Bush administrations to drive home the point that while the surplus savings to create the housing bubble came from China, and the low interest rates to encourage borrowing came from the Fed, the incentive to lend irresponsibly to sub-prime borrowers came from a combination of governments and pressure groups, far more than it came from alleged deregulation or from a new outbreak of 'greed'. And this was the biggest reason for the collapse of so many banks and the insurance giant AIG. To leave Fannie and Freddie out of the story of the Great Recession is impossible, and to omit the political mandates that drove them is unthinkable. They were from start to finish a top–down distortion of a bottom–up market. David Stockman in his book *The Great Deformation* is unsparing in his conclusion: 'The Fannie Mae saga demonstrates that once crony capitalism captures an arm of the state, its potential for cancerous growth is truly perilous.' Jeff Friedman, in a lengthy and influential essay on the financial crisis, came to a similar conclusion: 'The financial crisis was caused by the complex, constantly growing web of regulations designed to constrain and redirect modern capitalism.' Peter Wallison, a member of the government's Financial Crisis Inquiry Commission, said something similar: 'The financial crisis was not caused by weak or ineffective regulation. On the contrary, the financial crisis of 2008 was caused by government housing policies.' The sub-prime crisis was a creationist, not an evolutionary phenomenon.

The evolution of mobile money

The government monopoly of money leads not just to the suppression of innovation and experiment, not just to inflation and debasement, not just to financial crises, but to inequality too. As Dominic Frisby points out in his book *Life After the State*, opportunities in finance ripple outwards from the Treasury. The state spends money before it even exists; the privileged banks then get first access to newly minted money and can invest it before assets have increased in cost. By the time it reaches ordinary people, the money is worth less. This outward percolation is known as the Cantillon Effect – after Richard Cantillon, who noticed that the creation of paper money in the South Sea Bubble benefited those closest to the source first. Frisby argues that the process of money creation by an expansionary government effectively redistributes money from the poor to the rich. 'This is not the free market at work, but a gross, unintended economic distortion caused by the colossal government intervention.'

The strange obsession that politicians have with determining the price of one currency in terms of another, rather than letting such a price emerge, has always baffled me. Britain, in particular, has a long history of crises caused by the mispricing of exchange rates. In 1925, Winston Churchill, as Chancellor of the Exchequer, took Britain back onto the gold standard at the wrong price, precipitating a recession. In 1967, James Callaghan resisted too long before devaluing the pound. In 1992, Norman Lamont tried to cling to a fixed rate of exchange with the deutschemark. And of course, in 1999 the European Union devised a painful trap in the form of a common currency, which delivered unemployment, deep recession and debt to the countries of southern Europe. What's with this obsession? Why can we not learn that prices cannot be fixed correctly by politicians? We do not set the price of toothpaste centrally, so why do we set the price of money so? Frisby again: 'This system of money and finance is not an unregulated free market, but protected crony capitalism. It is immoral, deeply unfair and highly perilous. It is exploited by rent-seekers.'

It is vital that the government's monopoly on the creation of money be broken. If, as US Congressman Ron Paul has argued, government is so sure that its own money is the best money, it should not fear competition: 'In a free market, the government's fiat dollar should compete with alternate currencies for the benefit of American consumers, savers and investors.' Having the right to opt out of the Bank of England's monopoly, says the British MP Douglas Carswell, 'might encourage it to stop taking liberties with our currency'.

Today, new forms of self-organising money are continually being born: air miles, mobile-phone credits, bitcoins. Will they eventually displace official currencies? I suspect they will. Kenya, unexpectedly, has led the way in developing mobile-phone money. In the early 2000s, unprompted by anybody in government or industry, Kenyans began transferring mobile-phone minutes to each other by text as a form of money. Mobile-phone operators like Safaricom and Vodafone realised what was happening and set out to ease the experience of users. M-Pesa now allows people to pay real money into their phones or take it out via agents, and to transfer credits between phones. This proved popular with people working in cities remitting cash to their families back home in rural villages. Two-thirds of Kenyans now use M-Pesa as money, and more than 40 per cent of the country's GDP flows through the currency. Far more Kenyans have access to financial saving and payment systems through their mobile phones than through conventional bank accounts.

A key ingredient in the success of the system in Kenya was that the regulator was kept out of the way, allowing the system to evolve. Not for want of trying: the banks have lobbied politicians to subject M-Pesa to more regulation. Elsewhere in the world, heavy-handed regulation stifled mobile money at birth. During Kenya's post-election violence in 2008, mobile-phone balances seemed a lot safer than cash, so the system gained further popularity. Soon it reached the critical mass where enough people were using M-Pesa that it made sense to join them, so as to be able to transact business with them. In Kenya people pay wages, purchase savings products and take out loans with M-Pesa cash.

Money serves three main functions – a store of value, a medium of exchange and a unit of account. These are often in conflict: gold works well as a store of value, being scarce and non-rusting; but it is too scarce to be a practical medium of exchange. Cowrie shells once served as a form of money in some parts of the world, because they are so hard and so rare. The problem with commodity money is that it is vulnerable to inflation if the supply suddenly increases – the discovery of a new source of cowries, or a new goldmine. Conversely, an alternative use for the commodity used as money can suddenly create a money shortage. When the Royal Navy started sheathing its hulls in copper, the price of copper rose to the point where people started melting down pennies for their more valuable copper content.

'Fiat' money, made of paper, say, avoids these problems, but since the only check on supply is the state's promise not to print money at whim, and since that promise has been broken not just once but repeatedly throughout history by states doing just that in order to reduce their debts, the search for a way to write rules of monetary policy that will not be broken continues. As the monetary economist George Selgin and colleagues have argued, on any objective measure, the first century of the US Federal Reserve's existence has been a failure. Not only has there been incontinent inflation since 1913, the year the Fed came into existence (8 per cent in the preceding 120 years, 2,300 per cent in the succeeding hundred years), but there has been devastating deflation too, and more banking panics, more financial volatility, longer and deeper recessions. Even the Fed's response to the crisis of 2008 has come under severe criticism, as it effectively bailed out bad assets while doing little to help solvent institutions with needed liquidity – the reverse of Walter Bagehot's lender-of-last-resort recommendation. Some think that the Fed turned a relatively modest economic recession caused by deflating house prices into a Great Recession by this bungled response. All in all, it is possible that future generations will conclude that the Fed has been to the economy as bleeding was to eighteenth-century medicine: worse than useless, but none dared say so. Wise men no more know how to centrally plan a monetary system than they know how to centrally plan factories, hospitals and railways.

An alternative monetary approach would be to find a form of 'synthetic commodity' money that would have no other use, so was not suddenly in demand elsewhere, but would have an immovable scarcity factor, so could be counted on to retain its value. Printing paper money but then ostentatiously destroying the lithograph, it used to be argued in the pre-computer days, would to some degree serve this purpose. In a similar vein, in Iraq in the 1980s Saddam Hussein issued dinar notes printed in Britain and engraved in Switzerland. After the first Gulf War, sanctions cut him off from the supply of his currency. He started printing money in Iraq, but the quality was poor, counterfeiting was easy and the quantity was too high, causing inflation. However, the Swiss-made dinars remained in circulation, and began to diverge in value from the local ones. Since there were no more being made, people saw them as a store of value and they held their value against the dollar.

And then came bitcoins. The implications of crypto-currencies, and their recent evolution, are profound; they go well beyond the subject of money. They give us a glimpse of the future evolution of the internet itself.

16

The Evolution of the Internet

Nothing can be made from nothing – once we see that's so,
Already we are on the way to what we want to know:
What can things be fashioned from? And how is it without
The machinations of the gods, all things can come about?

Lucretius, *De Rerum Natura*, Book 1, lines 164–7

The internet has no centre and no hierarchy. All the computers that use it are equal – 'peers' in a network. As Steven Berlin Johnson remarks, the internet is not even a bottom–up system, for the existence of a bottom implies a top. And nobody planned it. Though it is the sum – the multiple, actually – of many individually deliberate projects, the internet as a whole has emerged in my lifetime, undesigned, unexpected, unpredicted. Nobody foresaw blogs, social networks, even search engines, in advance, let alone the particular forms they took. Nobody is in charge. Yet for all its messiness, the internet is not chaotic. It is ordered, complex and patterned. It is a living example, before our eyes, of the phenomenon of evolutionary emergence – of complexity and order spontaneously created in a decentralised fashion without a designer.

It is worth recalling just how pessimistic most people were

about communication technology during the twentieth century. George Orwell saw brainwashing as the future of radio and television. In his book *The Constitution of Liberty,* Friedrich Hayek thought we were 'only at the threshold of an age in which the technological possibilities of mind control are likely to grow rapidly'.

Indeed, in the early part of the twentieth century, when the only mass-communication technologies were radio and films, power shifted towards totalitarians in short order. These technologies were suited to being broadcast from one to many. Christopher Kedzie of Harvard points out that dictators like communication technologies that have very few originators and very many recipients. Many-to-many technologies, like the telephone and the internet, have undermined rather than strengthened dictatorial government. It is no accident that in East Germany in 1988, 52 per cent of households owned a colour television, while just 4 per cent owned a telephone. Few can doubt that the internet is a force for liberty of the individual.

There is a long and sterile argument to be had about who deserves credit for inventing the internet – government or private industry. Barack Obama is in no doubt that, as he put it in a speech in 2012, 'The Internet didn't get invented on its own. Government research created the Internet.' He was referring to the fact that the decentralised network we know today began life as the Arpanet, a project funded by the Pentagon, and that relied on an idea called packet switching, dreamt up by Paul Baran at the RAND Corporation, whose motive was chiefly to make something that could survive a Soviet first strike and still transmit messages to missile bases to retaliate. Hence the decentralised nature of the network.

That's nonsense, say others. The internet is more than package-switching. It requires computers, communications, all sorts of software and other protocols, many of which the government-funded research projects would have bought from private enterprise. Anyway, if you really want to see the Arpanet as the origin of the internet, please explain why the government sat on it for thirty years and did almost nothing with it until it was effectively privatised in the 1990s, with explosive results. Indeed, it's worse than that. Until 1989 the government actually prohibited the use of the Arpanet for private or commercial purposes. A handbook for users of the Arpanet at MIT in the 1980s reminded them that 'sending electronic messages over the ARPAnet for commercial profit or political purposes is both antisocial and illegal'. The internet revolution might have happened ten years earlier if academics had not been dependent on a government network antipathetic to commercial use.

Well, then, perhaps we should forget about who was funding the work, and at least give credit to the individuals without whom the internet would never have happened. Paul Baran was first with the notion of packet switching, Vint Cerf invented the TCP/IP protocols that proved crucial to allowing different programs to run on the internet, and Sir Tim Berners Lee developed the worldwide web. Yet there is a problem here, too. Can anybody really think that these things – or their equivalents – would not have come into existence in the 1990s if these undoubtedly brilliant men had never been born? Given all we know about the ubiquitous phenomenon of simultaneous invention, and the inevitability of the next step in innovation once a technology is ripe (see Chapter 7), it is inconceivable that the twentieth century would have ended without a general, open means of connecting computers to each other so that people could see what was on other nodes than their own hard drive. Indeed, the notion of packet switching – and even the name we now use for it – occurred independently to a Welshman named Donald Davies just a short time after Baran stumbled on it. Vint Cerf shares the credit for TCP/IP with Bob Kahn. So, while we should honour individuals for their contributions, we should not really think that they made something come into existence that would not have otherwise. The names would be different, and some of the procedures too, but an alternative internet would exist today whoever had lived.

The true origin of the internet does not lie in brilliant individuals, nor in private companies, nor in government funding. It lies, as Steven Berlin Johnson has argued persuasively, in open-source, peer-to-peer networking of an almost hippie, sixties-California-commune kind. 'Like many of the bedrock technologies that have come to define the digital age, the internet was created by – and continues to be shaped by – decentralised groups of scientists and programmers and hobbyists (and more than a few entrepreneurs) freely sharing the fruits of their intellectual labor with the entire world.' These were people collaborating because they wanted to, not because they were paid to, and with little or no intellectual property in their ideas. Open-source collaborative networks created a huge proportion of the lines of code on which the internet depends today – and not just the internet, but smartphones, stock markets and aeroplanes. The operating system of the computer I am writing this on is based on the UNIX operating system, something that was built by collaboration, but not for profit. The web servers that I am using to research my facts are powered by Apache software, another open-source program. This is, to borrow John Barlow's phrase, 'dot-communism': a sharing, swapping community of people who contribute to joint effort and expect no private rewards. What a splendid irony, that from the bowels of the Cold War military-industrial complex in the capitalist United States, there emerged a technology of 'dense, diverse and decentralised exchange' that is producing something far more like the ideal of Marxism than communist regimes ever did.

The balkanisation of the web

For a while, we all got the point. We crowd-sourced and wiki-ed and clouded our lives. Journalists, those most anarchic of beasts, found themselves overtaken by bloggers, tweeters and amateur cameramen, and they did not like it. Only top–down journalism could do proper investigation, they said. Scientists had to get used to irreverent and instantaneous discussion of their ideas on forums, rather than stately and opaque clubs of peer review and publication. Politicians had to put up with abuse on Twitter.

But then the fightback began. What the columnist Matthew Parris calls the snoopers, censors and web wardens began to proliferate. In Cuba and China they kept the internet opaque, but in other countries too they gnawed away at freedom. We learned in recent years that America's security state, just as much as Russia's and China's, is hell-bent on spying electronically on its citizens, then lying about the fact, while justifying its actions with secret interpretations of the law. The communications revolution was being used, in Eben Moglen's words, to 'fasten the procedures of totalitarianism on the substance of demo-cratic society'. The governments of America, Europe and Asia, it emerged, all implicitly agreed that they should be free to listen to each other's populations' conversations. Only nobody told those populations that this was the new agreement.

It was a pity, perhaps, that we found all this out from flawed whistleblowers like Julian Assange and Edward Snowden, who sometimes seemed only too happy to compound the state's sins by then exposing the contents of the eavesdropping themselves (and throwing themselves on the mercy of illiberal regimes). But you can disapprove of state snooping without approving of the leaks of the snoops. If ever anybody thought that the collapse of communism in 1989 would reduce the need for Western governments to behave secretly and illiberally, they have now been cruelly disillusioned. The very governments that wish to regulate what we do on the internet wish to be free to invade our privacy. In Britain, as Snowden revealed, over a million web-cam users were spied upon in a fishing expedition by the govern-ment spying agency, GCHQ – undertaken with no excuse of a suspicion of wrongdoing.

The authoritarians surely won't win, but they will succeed in turning parts of the system into top–down fiefs. From the moment of the internet's birth, the usual suspects have been demanding a framework, an authority, a little bit of '*ordnung*'. A key battle in this war was the Stop Online Piracy Act introduced into Congress in 2011 at the behest of big Hollywood studios and other media companies reliant on intellectual property. With bipartisan support and much encouraged by the bureaucracy of big government, which remains horrified by the anarchy of the internet, the Bill looked certain to pass. But an unexpected last-minute rebellion in January 2012, when hundreds of websites went black in protest against the proposed law, killed it within a week.

The war is not over, however. Even organisations like Wikipedia succumbed to the authoritarian twitch, appointing editors with special privileges who could impose their own prejudices upon certain topics. The motive was understandable – to stop entries being taken over by obsessive nutters with weird views. But of course what happened, just as in the French and Russian revolutions, was that the nutters got on the committee. The way to become an editor was simply to edit lots of pages, and thereby gain brownie points. Some of the editors turned into ruthlessly partisan dogmatists, and the value of a crowd-sourced encyclopedia was gradually damaged. As one commentator puts it, Wikipedia is 'run by cliquish, censorious editors and open to pranks and vandalism'. It is still a great first port of call on any uncontroversial topic, but I find Wikipedia cannot be trusted on many subjects. An entirely fictional war in the Indian state of Goa was invented, and not only survived for five years on Wikipedia but became a popular entry and won an award.

A small example, maybe, but one of many instances in recent years to show how Wikipedia had moved away from being a crowd-sourced thing to something more hierarchical and centrally controlled. Meanwhile, professional public relations firms do a lot of work to bias Wikipedia, and the net in general, in favour of their clients. A decision by the Court of Justice of the European Union in 2014 – that people should be allowed to insist on the deletion from search results of old stories about themselves, even if these were true – was a gift for crooks of all kinds.

And then there's real censorship, of the kind done by the Chinese state in particular. The number of countries that censor the internet has grown steadily, and now stands at more than forty. The tradition of what Vint Cerf calls 'permissionless innovation' is crucial to the success of the internet, and is under explicit attack from governments and busybodies all around the world who insist that all innovation must seek permission. The International Telecommunications Union, a United Nations body with 193 members, has been lobbied by several governments to extend its control over the internet, grab power over the registration of domain names and bring in international rules banning, for instance, the use of anonymity. While there are plenty of us who would like to see abusive internet commentators stripped of their anonymity, so would the leaders of repressive regimes like to see dissidents exposed. Russian President Vladimir Putin has been explicit that his goal is 'establishing international control over the Internet' through the ITU. In 2011 Russia joined with China, Tajikistan and Uzbekistan to propose an 'International Code of Conduct for Information Security' to the UN General Assembly.

The issue came to a head at a meeting of the ITU in Dubai in December 2012, where member countries voted by eighty-nine to fifty-five to give the United Nations agency unprecedented power over the internet, with Russia, China, Saudi Arabia, Algeria and Iran leading the charge for regulation. Even though many countries refused to sign the new treaty, the head of America's Federal Communications Commission argued that serious damage had still been done to free speech around the world, because pro-regulation forces had already succeeded in changing the meaning of crucial treaty definitions that were understood to insulate the internet from intergovernmental control. He said that the ITU's 'appetite for regulatory expansionism is insatiable'.

For all its decentralised nature, the internet does have a central committee – the Internet Corporation for Assigned Names and Numbers, or ICANN. The American government set it up, though it now shares responsibility with other governments and international bodies. This firm has gleaming corporate offices and the power to hand out domain names.

In general I remain optimistic that the forces of evolution will outwit the forces of command and control, and the internet will continue to provide a free space for all. But only because of human ingenuity staying one step ahead of the *dirigistes*. Perhaps the most profoundly important of the internet's offspring will be digital currencies independent of government: bitcoin, or the crypto-currencies that will come after it. 'I think that the Internet is going to be one of the major forces for reducing the role of government. The one thing that's missing, but that will soon be developed, is a reliable e-cash,' said Milton Friedman. And it is not just e-cash; it is the technology behind bitcoin that could finally decentralise not just the internet but society too. The blockchain technology that makes bitcoin work has far-reaching implications.

The bizarre evolution of blockchains

The story begins in 1992, when the internet was just beginning to emerge. A wealthy computer pioneer named Tim May invited a group of people to his house in Santa Cruz to discuss how to use 'cryptologic methods' on networked computers to break down barriers of intellectual property and government secrecy. 'Arise! You have nothing to lose but your barbed wire fences,' he told them. They called themselves the 'cypherpunks', and they foresaw the way technology was both a threat to and an opportunity for freedom: a chance to open up the world, but a chance for the state to invade our lives. Their manifesto declared: 'We the Cypherpunks are dedicated to building anonymous systems. We are defending our privacy with cryptography with anonymous mail forwarding systems, with digital signatures, and with electronic money.'

Like most libertarian collectives, the cypherpunks' web community soon broke up in acrimonious bickering and flame wars. But not before they had sparked some interesting thoughts in each other's heads. The key names in this group were Adam Back, Hal Finney, Wei Dai and Nick Szabo. In grappling with the problems of anonymous, self-organising money systems, Back invented a system called hashcash, Dai came up with b-money, and Finney developed a vital protocol called 'reusable proofs of work'. It was Szabo who went furthest into the history and philosophy of the topic. With a degree in computer science and a doctorate in law, he became fascinated by the history of money, writing a lengthy essay on the subject, in which he explored a throwaway remark by the evolutionary biologist Richard Dawkins that 'money is a formal token of delayed reciprocal altruism' – or money makes it possible to pay back favours indirectly and at any time.

This essay, entitled 'Shelling Out: The Origins of Money', showed a keen appreciation of the fact that money evolved gradually and inexorably, and not by design. Money began with collectibles – items like shells, bones and beads, valued for their lack of perishability – which early human beings are known to have gathered, then gradually came into its other role as a medium of exchange, so that barter could be generalised. Szabo showed special interest in his essay in the ideas of evolutionary psychology, citing many works on the topic. By the 2000s he was musing about something called bitgold, an imaginary software product that would mimic the properties of gold: it would be scarce and hard to acquire, but easy for others to verify, and thus could be trusted as a store of value. Clearly, he was trying to think how to recreate online the key steps in the evolution of real money.

Some years went by. Then, on 18 August 2008, a month before the financial crisis broke in earnest, a new domain name was registered anonymously: bitcoin.org. Two weeks later, somebody with the user name 'Satoshi Nakamoto' posted a nine-page paper outlining an idea for a peer-to-peer electronic cash system called bitcoin. The bitcoin system went live a few months later, on the day the British government reported its second bailout of the banks, an event referred to by Satoshi, who quoted a headline from *The Times* in his announcement of bitcoin's birth. A month later Satoshi announced on the Peer-to-Peer Foundation website: 'I've developed a new open source P2P e-cash system called Bitcoin. It's completely decentralised, with no central server or trusted parties, because everything is based on crypto proof instead of trust. Give it a try, or take a look at the screenshots and design paper.' His motivation was clear. Bitcoin was designed to maintain its value without any precious-metal backing, without any centralised issuer, and without any intrinsic value. Satoshi invited users to 'escape the arbitrary inflation risk of centrally managed currencies!'

It is hard to get your head around how bitcoin works. One of the pithiest explanations I have come across is in a recent launch by Ethereum, a business built to follow up on bitcoin: 'The innovation provided by Satoshi is the idea of combining a very simple decentralised consensus protocol, based on nodes combining transactions into a "block" every ten minutes, creating an ever-growing blockchain, with proof of work as a mechanism through which nodes gain the right to participate in the system.' If you think that's hard to understand, you are not alone. I have yet to come across a description of blockchain technology in English, as opposed to mathematics, that is really clear. In outline, I know that bitcoin is effectively a public ledger – a compendium of transactions, stored by bitcoin users all over the world. To participate, you effectively create a part of that ledger, and share it with others as a cryptographically bound 'block'. This makes bitcoin infallible and public as a register of who has transferred value to whom, with no bank or other body verifying the fact.

Satoshi Nakamoto is a pseudonym. The founder or founders of bitcoin wished to remain anonymous, for fairly obvious reasons. Previous inventors of private money had often ended up in deep trouble with a jealous state. Bernard von NotHaus, for example, starting in 1998, quite openly minted and sold coins called 'liberty dollars' made from gold, with absolutely no pretence that they were fake dollars. He set out to compete with the Federal Reserve in the same way that Federal Express competes with the Post Office: offering an alternative store of value. After eleven years of tolerating this, suddenly and without warning the United States federal government raided, arrested and prosecuted him for counterfeiting, fraud and conspiracy against the United States. Despite the fact that his customers were neither deceived nor dissatisfied, he was convicted – effectively of competing against the federal government. Then there was e-gold, a digital payments system run from the Caribbean by an oncologist called Doug Jackson that rocketed to $1.5 billion in transactions before being shut down on the grounds that it was allowing illegal money transmission. Governments do not take kindly to money that is outside their control. Hence the shyness of bitcoin's founder.

The mysterious founder

Who is Satoshi Nakamoto? *Newsweek* magazine thought it had found him in March 2014 when it identified a sixty-four-year-old Japanese-American programmer named Dorian Satoshi Nakamoto living near Los Angeles. The baffled and beleaguered Dorian, an unemployed man in poor health with a clumsy command of English, protested that he had nothing to do with bitcoin, did not understand what it was, and thought it was called 'bitcom'. And, he asked pertinently, why would he use part of his real name if he wanted to stay anonymous? Satoshi himself emerged briefly from seclusion to announce on the web (anonymously) that he was not Dorian.

The 'real' Satoshi uses a Japanese name, a German web address, lots of British phrases and references, and, judging by the timing of his posts, keeps American (east coast) hours. The only high-tech region he does not seem to be associated with in any way is the west coast of North America, where Nick Szabo lives. Forensic analysis of his style, his idiosyncrasies, his likely age and the pattern of his activity has led the author Dominic Frisby and others – including a team of forty forensic linguists from Birmingham University – to the conclusion that Satoshi Nakamoto is probably Nick Szabo. Suspiciously, the normally prolific Szabo went unusually silent around the time Satoshi Nakamoto became active, and vice versa. However, Szabo has denied on Twitter that he is Satoshi. (Some still think that he and Hal Finney collaborated as Satoshi, giving each deniability.) Szabo himself keeps a low profile. No photograph of him can be found on the net.

Whoever he is, 'Satoshi Nakamoto' knows a lot about computer programming and economic history – a rare combination. There is little doubt that bitcoin is one of the most significant inventions of our lifetime (though I doubt it would have remained uninvented if Satoshi had not existed: someone else would have come up with some form of self-verifying currency). Bill Gates calls it a tour de force. So far it has proved impossible to hack, it has characteristics that make it almost ideal as a system of money, it is self-policing, it is impossible to inflate, and it is beyond the reach of the state. It solves the problem that bedevilled all previous forms of electronic money: that you need a third party to ensure that the money somebody is sending you is not being sent to somebody else at the same time. That's what a bank does with money transfers, and a government does by minting limited numbers of coins and notes. Bitcoin prevents double spending by making sure that if the same money is sent to two places, only the transaction that is confirmed first will process.

The minting of bitcoins mimics the act of mining: it was easy at first, but has gradually become much more difficult, so that now huge banks of computers are required to mine each coin. Each coin consists of a chain of previously mined codes, called a blockchain, plus one new block, which is created by the solving of a difficult puzzle by hard computer grind. At the time of writing, about thirteen million bitcoins are in circulation, and the number can never exceed twenty-one million. The rate of production halves every four years, until the total supply tops out in the middle of the twenty-second century.

You can buy or sell a bitcoin as you can a pound or a dollar. The price shot up in the wake of the financial crisis in Cyprus in 2013, when private depositors woke up to the fact that their conventional money was not safe in banks, because the government of Cyprus announced that it would seize over 40 per cent of all savings over $100,000. As investors around the world digested the arbitrary power of governments, bitcoin's price rose from about $120 in September 2013 to almost $1,200 in December of that year. It has since slowly declined.

At the time of writing, about $6 billion worth of money is held in bitcoins. But it is still a long way from taking over as the world's reserve currency. It does not yet work as a unit of account. The volatility and bubble-like behaviour of bitcoins are not encouraging for a world reserve currency, and nor is its relatively small supply. It is also still not easy to get many traders, even online, to accept bitcoins. The first bitcoin exchange, Mt. Gox, collapsed in a pile of fraud. Moreover, bitcoins have proved very popular with drug dealers, especially via an online exchange called Silk Road. The authorities have infiltrated Silk Road and busted a number of criminals (including a twenty-nine-year-old von-Mises-quoting dropout calling himself Dread Pirate Roberts, who operated from a coffee shop in San Francisco). All these factors have tarnished the electronic ledger's reputation.

So don't hold your breath, or conclude that bitcoin is the final future of money. It is more like the beginning of something. And there is no doubt that crypto-currencies will evolve. As Kevin Dowd, a professor of finance at Durham University, points out in relation to Silk Road, 'Each bust works as evolutionary pressure, weeding out the weaker sites and teaching the others what to avoid. Cut one head off, and new ones will take its place: Silk Road 2.0 is already up and operating.'

As Dominic Frisby remarks, not only has bitcoin's evolution so far been chaotic, unplanned and organic, but the people around it are 'an eclectic mix of all sorts from the computer whizz to the con artist to the economist; from the opportunist to the altruist to the activist'. None the less, it is worth remarking just how much the humble bitcoin has achieved in a world where it has no intrinsic value whatsoever, which bodes well for future crypto-currencies online. There are now more than three hundred rival online crypto-currencies competing with bitcoins – altcoins, they are called – and though none has yet gained anything like the market share of bitcoin, it may only be a matter of time.

Just imagine what might happen if decentralised crypto-currencies really do take off. If people started putting their savings in them, and financial firms started offering interesting crypto-currency-based products, governments would find their room for manoeuvre much diminished. They could not borrow profligately, or tax rapaciously, or spend freely without looking over their shoulders to see what it might do to their currency against (say) bitcoin. Frisby thinks it would force the state to tax consumption rather than production, and it would drive inflation out of the system. Above all, it would put the big banks out of business, removing the distortion by which so much of the world's wealth has ended up concentrated in one industry. Satoshi Nakamoto says bitcoin is 'very attractive to the libertarian viewpoint if we can explain it properly'. Nassim Taleb says, 'Bitcoin is the beginning of something great: a currency without a government, something necessary and imperative.' Kevin Dowd says it 'raises profound issues of an emerging spontaneous social order . . . a crypto-anarchic society in which there is no longer any government role in the monetary system'. Jeff Garzik, a bitcoin developer, calls it 'the biggest thing since the internet – a catalyst for change in all areas of our lives'.

Blockchains for all

What are these enthusiasts on about? The 'blockchain' technology behind bitcoin could prove to be an ingredient of an entire new world of technology, as big as the internet itself, a wave of innovation that drives the middleman out of much commerce and leaves us much more free to exchange goods and services with people all over the world without going through corporate intermediaries. It could radically decentralise society itself, getting rid of the need for banks, governments, even companies and politicians.

Take the example of Twister, a blockchain-based rival to Twitter, built entirely on a peer-to-peer network. If you live under a despotic regime, sending a message critical of your government on Twitter leaves you vulnerable to that government coercing Twitter, the company, into handing over your details. With Twister, that will not be possible. Then there is Namecoin, which aims to issue internet names in a decentralised, peer-to-peer fashion; Storj, which plans to allow cloud storage of files hidden inside blockchains; and Ethereum, which is a decentralised peer-to-peer network 'designed to replace absolutely anything that can be described in code', as Matthew Sparkes puts it. The digital expert Primavera De Filippi sees Ethereum and its ilk coming up with smart contracts, allowing 'distributed autonomous organisations' that, once they have been deployed on the blockchain, 'no longer need (nor heed) their creators'.

In other words, not just driverless cars, but ownerless firms. Imagine in the future summoning a taxi that not only has no driver, but that belongs to a computer network, not to a human being. That network has raised funds, signed contracts and taken delivery of vehicles, even though its 'headquarters' is distributed all over the net. That would represent the triumph of decentralised, evolving, autonomous systems. It would mean that 'software has achieved what regulation has failed to achieve', in the words of Andreas Antonopoulos of Blockchain.info. He argues that unlike centralised systems, decentralised institutions are resilient and incorruptible. 'There is no centre, they do not afford opportunities for corruption. I think that's a natural progression of humanity.'

You may think I am listening too credulously to radical libertarian dreamers, and perhaps I am. My confidence that something big is coming has its roots in the evidence that I have recounted in this book of the evolution of systems that are the result of human action, but not of human design. Something as radical as language or government is emerging from the internet. Officials, lawyers, politicians, businessmen may come together to try to stop this, glimpsing their own redundancy, and for a while they may succeed. But the inexorable, inevitable, implacable nature of evolution will eventually defeat them. Remember how technology evolves, whether we want it to or not.

Re-evolving politics

Take politics. Even today the internet revolution is undermining Leviathan at every turn. The internet turns everybody into a journalist and a politician; puts the customer in ultimate charge; and lowers the cost for ordinary people to do extraordinary things, whether in charity, business or politics. Big companies are tumbling before its creative-destructive onslaught; big state bureaucracies cannot long resist. As the maverick MP Douglas Carswell puts it, 'Everything that the internet touches it transforms. The barriers to entry come crashing down. Established operators face competition from nimble upstarts. So, too, in politics.' Carswell argues that i-democracy is rapidly and inexorably transforming the old ways of doing politics, replacing party-controlled, bureaucracy-enabled traditions with radical emergent possibilities, from open primaries to instant plebiscites, from participatory budgeting in local government to online recall. 'It has awakened something magnificently Cromwellian in our democracy.' The big-government model that threatens to bankrupt and bully us is not just unaffordable; it is also increasingly impractical. In a world where individuals and firms can hop easily between jurisdictions – where geese do not hang around waiting to be plucked by the taxman – it will be increasingly hard to justify wasteful extravagance in the public finances. And that will be doubly true if crypto-currencies become widely available.

Carswell envisages a world in which you, the citizen, are in charge. The official who dictated a one-size-fits-all policy has to do as you tell him; so too does the elected politician who in the past took his instruction from you just once every four or five years. Says Carswell: 'The digital revolution is a coup d'état against the tyranny of this elite. It overthrows these second-hand dealers in other people's ideas.' When the Conservative Daniel Hannan stood up in the European Parliament in 2009 and lambasted a hapless Gordon Brown, the British Prime Minister, for three minutes, the mainstream media at first ignored it. But within minutes the speech had gone viral on YouTube, and when it had been viewed more than a million times, the mainstream media was obliged to catch up. Revealingly, the editor of the *New Statesman*, Peter Wilby, said the episode showed how lacking in quality control the internet was – by which he meant unfiltered by people like him. It is filtered now by collective wisdom.

Carswell points out that politics has become steadily more centralised in recent decades, but he thinks he detects the beginning of the reversal of this trend. The state has grabbed more and more of the money made in a country, and spent it on designing solutions at the centre, on political creationism. It has emasculated elected representatives by transferring powers to unelected officials. Four-fifths of legislation in Britain is now authored by the unelected and permanent civil service, whose job has changed from implementing to making policy. The elected officials who do have influence constitute a small group of courtiers around the head of government, and in the 1990s they perfected the tight, centralised control of policy and politics embodied in the person of the spin doctor. And the political system, with its bias in favour of the status quo, its precautionary mistrust of the new, and its elite assumptions, is almost perfectly designed to frustrate all attempts at innovation.

But this is changing fast. The traditional political parties no longer meet people's political needs. The state treats them far worse than business does. People who have better and better experiences as citizens – being able to change suppliers, demand decent service, get instant information online, purchase shoes with a single click – are increasingly frustrated that they are treated so badly as subjects of a government. Why must queries take weeks to be answered? Why must websites be so patronising? Why must forms be so badly designed? Why must fees for service be so opaque? Why must legislation be so inflexible? The opportunity presented by the digital revolution to 'hyper-personalise' public services is huge. Put parents in charge of their children's individual education budget; patients in charge of their own health budget; cut out the bureaucratic middleman.

Digital democracy could shake up government as radically as the end of the Cold War shook up communism. To date the impact of digital technologies on the practice and productivity of government has been almost non-existent. If anything, productivity in public services has gone down, not up. That is a truly staggering statistic, when you think about it. Computers and smartphones and dirt-cheap communication and the infinite resources of the internet have arrived in their offices, yet bureaucrats have managed to avoid nudging their productivity up at all? How did they manage that? An earthquake is coming. Let politics evolve.

EPILOGUE

The Evolution of the Future

There are two ways to tell the story of the twentieth century. You can describe a series of wars, revolutions, crises, epidemics, financial calamities. Or you can point to the gentle but inexorable rise in the quality of life of almost everybody on the planet: the swelling of income, the conquest of disease, the disappearance of parasites, the retreat of want, the increasing persistence of peace, the lengthening of life, the advances in technology. I wrote a whole book about the latter story, and wondered why it seemed original and surprising to do so. It was surely gloriously obvious that the world was a much, much better place than it had ever been. Yet read the newspapers and you would think we had lurched from disaster to disaster, and faced a future of inevitable further disaster. Glance at school history curriculums and you find them utterly dominated by the disasters of the past – and the crises of the future. I could not quite reconcile in my mind this strange juxtaposition of optimism and pessimism. In a world that delivers an endless supply of bad news, people's lives get better and better.

Now I think I understand, and it has been the purpose of this book partly to explore that understanding. To put my explanation in its boldest and most surprising form: bad news is man-made, top–down, purposed stuff, imposed on history. Good news is accidental, unplanned, emergent stuff that gradually evolves.

The things that go well are largely unintended; the things that go badly are largely intended. Let me give you two lists. First: the First World War, the Russian Revolution, the Versailles Treaty, the Great Depression, the Nazi regime, the Second World War, the Chinese Revolution, the 2008 financial crisis: every single one was the result of top–down decision-making by relatively small numbers of people trying to implement deliberate plans – politicians, central bankers, revolutionaries and so on. Second: the growth of global income; the disappearance of infectious diseases; the feeding of seven billion; the clean-up of rivers and air; the reforestation of much of the rich world; the internet; the use of mobile-phone credits as banking; the use of genetic finger-printing to convict criminals and acquit the innocent. Every single one of these was a serendipitous, unexpected phenomenon supplied by millions of people who did not intend to cause these big changes. All the interesting things are incremental, says the psephologist Sir David Butler, and very few of the major changes in the statistics of human living standards of the past fifty years were the result of government action.

Of course, you can find counter-examples: where one individual or institution did an especially good thing according to a plan (the moon landings?); or one emergent phenomenon was disastrously bad (the rise of allergies and auto-immune disorders as a result of excessive hygiene?). But I submit that there are not nearly as many of these. Letting good evolve, while doing bad, has been the dominant theme of history. That is why the news is full of only bad things being done, but we find when they are over that great good has happened unheralded. Good things are gradual; bad things are sudden. Above all, good things evolve.

But surely, I hear you cry, this is an absurd exaggeration. The world is chock full of designed, planned and intended things that work fine. Yet just because something is ordered does not mean it was designed. As often as not it emerged through seren-dipitous trial and error. Equating order with control retains a powerful intuitive appeal, Brink Lindsey has pointed out.

'Despite the obvious successes of unplanned markets, despite the spectacular rise of the Internet's decentralized order, and despite the well-publicized new science of "complexity" and its study of self-organizing systems, it is still widely assumed that the only alternative to central authority is chaos.'

Even the paradigmatic examples of beautiful design – such as the wonderful Macbook Air laptop on which I write these words – are actually the result of an evolutionary process, which not only combined the work of thousands of inventors, but winnowed through myriad possible designs and selected this version before placing it in front of the market to be selected or rejected. True, Sir Jonathan Ive gets the credit, and rightly so, for many of Apple's outstanding designs, including this one, but the ingredients and constituent parts – the silicon chips, the software, the anodised aluminium casing – owe their origins to other inventors. The process that combined and selected them was bottom–up. This laptop evolved at least as much as it was created.

As I argued in the prologue, the theory of evolution by natural selection as outlined by Charles Darwin in 1859 should really be called the 'special theory' of evolution, to distinguish it from the 'general theory' of evolution. I owe this notion to Richard Webb, an expert on both evolution and innovation. The point he is making is one that I have tried to develop in this book, namely that the flywheel of history is incremental change through trial and error, with innovation driven by recombination, and that this pertains in far more kinds of things than merely those that have genes. This is also the main way that change comes about in morality, the economy, culture, language, technology, cities, firms, education, history, law, government, religion, money and society. For far too long we have underestimated the power of spontaneous, organic and constructive change driven from below, in our obsession with designing change from above. Embrace the general theory of evolution. Admit that everything evolves.

It is a fair bet that the twenty-first century will be dominated mostly by shocks of bad news, but will experience mostly invisible progress of good things. Incremental, inexorable, inevitable changes will bring us material and spiritual improvements that will make the lives of our grandchildren wealthier, healthier, happier, cleverer, cleaner, kinder, freer, more peaceful and more equal – almost entirely as a serendipitous by-product of cultural evolution. But the people with grand plans will cause pain and suffering along the way.

Let's give a bit less credit to creationists, while we encourage and celebrate the evolution of everything.

ACKNOWLEDGEMENTS

This book has been years, perhaps decades, in gestation, so it is impossible to thank everybody who has given me inspiration and food for thought over that time. As I am obsessed with pointing out, the whole point of human thought is that it is a distributed phenomenon, living between and among human brains, rather than inside them. I am just a node in a huge network of knowledge, trying to capture an ethereal and evolving entity in a few inadequate words. This is not to imply that anybody but myself deserves blame for any mistake in the book.

None the less, many people deserve particular thanks for being extremely generous with their thoughts, their suggestions, their warnings and their time. They include – but are not limited to – the following: Brian Arthur, Eric Beinhocker, Don Boudreaux, Karol Boudreaux, Giovanni Carrada, Douglas Carswell, Monika Cheney, Gregory Clark, Stephen Colarelli, John Constable, Patrick Cramer, Rupert Darwall, Richard Dawkins, Daniel Dennett, Megnad Desai, Kate Distin, Bernard Donoughue, Martin Durkin, Danny Finkelstein, David Fletcher, Bob Frank, Louis-Vincent Gave, Herb Gintis, Hannes Gissurarson, Dean Godson, Oliver Goodenough, Anthony Gottlieb, Brigitte Granville, Jonathan Haidt, Daniel Hannan, Tim Harford, Judith Rich Harris, Joe Henrich, Dominic Hobson, Tom Holland, Lydia Hopper, Anula Jayasuriya, Terence Kealey, Hyperion Knight, Kwasi Kwarteng,

Norman Lamont, Nigel Lawson, Kui Wai Li, Mark Littlewood, Niklaas Lundblad, Deirdre McCloskey, Geoffrey Miller, Alberto Mingardi, Sugata Mitra, Andrew Montford, Tim Montgomerie, Jon Moynihan, Jesse Norman, Selina O'Grady, Gerry Ohrstrom, Jim Otteson, Owen Paterson, Rose Paterson, Benny Peiser, Venki Ramakrishnan, Neil Record, Pete Richerson, Adam Ridley, Russ Roberts, Paul Romer, Paul Roossin, David Rose, George Selgin, Andrew Shuen, Emily Skarbek, Bill Stacey, John Tierney, Richard Tol, James Tooley, Andrew Torrance, Nigel Vinson, Andreas Wagner, Richard Webb, Linda Whetstone, David Sloan Wilson, John Witherow, Andrew Work, Tim Worstall, Chris Wright – and many more.

In researching and writing the book I had much valuable, practical help from Guy Bentley and Andrea Bradford. My sincere thanks to them. My agents, Felicity Bryan and Peter Ginsberg, and my editors, Louise Haines and Terry Karten, were patient, encouraging and acute throughout.

My greatest thanks go to my family, Anya, Matthew and Iris, who contributed not only ideas and insights, but sanctuary and sanity.

SOURCES AND FURTHER READING

Prologue: The General Theory of Evolution

On energy evolution, Bryce, Robert 2014. *Smaller Faster Lighter Denser Cheaper*. PublicAffairs.

On antifragility, Taleb, Nassim Nicholas 2012. *Antifragile*. Random House.

On Adam Smith, *The Theory of Moral Sentiments*. 1759.

On Adam Ferguson, *Essay on the History of Civil Society*. 1767.

On the lack of a name for objects that are the result of human action but not human design, Roberts, R. 2005. The reality of markets. At Econlib.org 5 September 2005.

Richard Webb's notion of a special and a general theory of evolution was enunciated during a Gruter Institute conference in London in July 2014.

Chapter 1: The Evolution of the Universe

On Lucretius, the translation I use here is a very lyrical one by the poet Alicia Stallings: Stallings, A.E. (translated and with notes) 2007. Lucretius. *The Nature of Things*. Penguin.

On skyhooks, Dennett, Daniel C. 1995. *Darwin's Dangerous Idea*. Simon & Schuster. The first use of the word is here: 'A naval aeroplane, with an officer pilot and a warrant or petty officer telegraphist, was cooperating with artillery in a new system of signalling. The day was cold and the wind was bumpy, and the aeroplane crew were frankly bored. Presently the battery signaller sent a message, "Battery out of action for an hour; remain aloft awaiting orders." Back came the reply with remarkable promptitude: "This machine is not fitted with sky-hooks." ' From the *Feilding Star* (New Zealand) 15 June 1915.

On the implications of Darwinism, Arnhart, Larry 2013. The Evolution of Darwinian Liberalism. Paper to the Mont Pelerin Society June 2013.

On Lucretius, Greenblatt, Stephen 2012. *The Swerve*. Vintage Books.

On Dawkins and Lucretius, Gottlieb, Anthony 2000. *The Dream of Reason*. Allen Lane/The Penguin Press.

On Lucretius's influence on Western thought, Wilson, Catherine 2008. *Epicureanism at the Origin of Modernity*. Oxford University Press.

On Newton and Lucretius, Jensen, W. 2011. Newton and Lucretius: some overlooked parallels. In T.J. Madigan, D.B. Suits (eds), *Lucretius: His Continuing Influence and Contemporary Relevance*. Graphic Arts Press; and Johnson, M. and Wilson, C. 2007. Lucretius and the History of Science. *The Cambridge Companion to Lucretius*, 131–148, ed. S. Gillespie and P. Hardie. Cambridge University Press.

On Newton's religious swerve, Shults, F.L. 2005. *Reforming the Doctrine of God*. Eerdmans Publishing.

On the swerve, Cashmore, Anthony R. 2010. The Lucretian Swerve: The biological basis of human behavior and the criminal justice system. *PNAS* 107:4499–4504.

On Voltaire and Lucretius, Baker, E. 2007. In *The Cambridge Companion to Lucretius*, 131–148, ed. S. Gillespie and P. Hardie. Cambridge University Press.

On Erasmus Darwin, Jackson, Noel 2009. Rhyme and Reason: Erasmus Darwin's romanticism. *Modern Language Quarterly* 70:2.

On Hutton, Dean, D.R. 1992. *James Hutton and the History of Geology*. Cornell University Press; and Gillispie, C.C. 1996. *Genesis and Geology*. Harvard University Press.

On determinism, Laplace, Pierre-Simon. 1814. *A Philosophical Essay on Probabilities*; what Laplace meant is discussed in Hawking, S. 1999. Does God Play Dice?. Public lecture, archived at archive.org; and Faye, Hervé 1884. *Sur l'origine du monde: théories cosmogoniques des anciens et des modernes*. Paris: Gauthier-Villars.

On the anthropic principle, Waltham, D. 2014. *Lucky Planet: Why the Earth is Exceptional and What That Means for Life in the Universe*. Icon Books.

Douglas Adams's puddle metaphor was in a speech in 1998. Quoted at biota.org/people/douglasadams/index.html.

On Voltaire and Emilie du Châtelet, Bodanis, David 2006. *Passionate Minds: The Great Enlightenment Love Affair*. Little, Brown.

Chapter 2: The Evolution of Morality

On Smith's moral philosophy, Macfarlane, Alan 2000. *The Riddle of the Modern World*. Palgrave; Otteson, James 2013. Adam Smith. In Roger Crisp (ed.), *Oxford Handbook of the History of Ethics*, 421–442. New York: Oxford University Press; Otteson, James 2013. *Adam Smith*. New York: Bloomsbury Academic; Otteson, James 1998. *Adam Smith's Marketplace of Life*. Cambridge University Press; Roberts, Russ 2005. The reality of markets. econlib.org/library/Columns/y2005/Robertsmarkets.html; and Roberts, Russ 2014. *How Adam Smith Can Change Your Life*. Penguin. Also Kennedy, G. 2013. Adam Smith on religion, in the *Oxford Handbook on Adam Smith*. Oxford University Press. And Foster, Peter 2014. *Why We Bite the Invisible Hand*. Pleasaunce Press. And Butler, Eamonn 2013. *Foundations of a Free Society*. IEA

On liberalism and evolution, Arnhart, Larry 2013. The Evolution of Darwinian Liberalism. Paper to the Mont Pelerin Society June 2013.

On the decline of violence, Pinker, Steven 2011. *The Better Angels of Our Nature*. Penguin.

On medieval violence, Tuchman, Barbara 1978. *A Distant Mirror*. Knopf.

On Lao Tzu, Blacksburg, A. 2013. Taoism and Libertarianism – From Lao Tzu to Murray Rothbard. Thehumanecondition.com.

On bourgeois values, McCloskey, Deirdre N. 2006. *The Bourgeois Virtues*. University of Chicago Press.

On Pope Francis, Tupy, Marion 2013. Is the Pope Right About the World?. *Atlantic Monthly* 11 December 2013.

On the common law, Hutchinson, Allan C. 2005. *Evolution and the Common Law*. Cambridge University Press; Williamson, Kevin D. 2013. *The End is Near and it's Going to be Awesome*. HarperCollins; Lee, Timothy B. 2009. The Common Law as a Bottom–Up System. Timothyblee.com 16 September 2009. And Hogue, Arthur R. 1966. *The Origins of the Common Law*. Indiana University Press. Also Hannan, Daniel 2012. Common Law, not EU Law. Xanthippas.com 20 March 2012. Also Boudreaux, Don 2014. Quotation of the Day 18 June 2014. At cafehayek.com.

On the evolution of law, Goodenough, Oliver 2011. When stuff happens isn't enough: how an evolutionary theory of doctrinal and legal system development can enrich comparative legal studies. *Review of Law and Economics* 7:805–820.

Chapter 3: The Evolution of Life

On Darwin and Adam Smith, Gould, Stephen Jay 1980. *The Panda's Thumb*. Norton; and Shermer, Michael 2007. *The Mind of the Market*. Times Books.

On natural theology, Paley, William 1809. *Natural theology; Or, evidences of the existence and attributes of the deity, collected from the appearances of nature*. London. Also Shapiro, A.R. 2009. William Paley's Lost 'Intelligent Design'. *Hist. Phil. Life Sci.* 31:55–78.

On the philosophy of Darwinism, Dennett, Daniel C. 1995. *Darwin's Dangerous Idea*. Simon & Schuster. And Cosmides, Leda and Tooby, John 2011. Origins of specificity. commonsenseatheism.com.

On Beverley's critique, Beverley, Robert Mackenzie 1867. *The Darwinian Theory of the Transmutation of Species*. James Nisbet & Co.

On pencils, 'I, Pencil' is by Leonard Reed (1958) and is easily accessed on the internet.

On Mount Improbable, Dawkins, Richard 1996. *Climbing Mount Improbable*. Norton.

On opsins, Feuda, R., Hamilton, S.C., McInerney, J.O. and Pisani, D. 2012. Metazoan opsin evolution reveals a simple route to animal vision. *Proceedings of the National Academy of Sciences*.

On redundancy in metabolic networks, Wagner, Andreas 2014. *Arrival of the Fittest*. Current Books.

On Kitzmiller vs Dover Area School District, 'Decision of the Court' is at talkorigins.org/faqs/dover/kitzmiller_v_dover_decision2.htm.

On Empedocles, Gottlieb, Anthony 2000. *The Dream of Reason*. Allen Lane/The Penguin Press.

On Harun Yahya, Tremblay, F. in 'An Invitation to Dogmatism'. At strongatheism.net.

On Gould's swerve, Dennett, Daniel C. 1995. *Darwin's Dangerous Idea*. Simon & Schuster.

On Wallace, Wallace, Alfred Russel 1889. *Darwinism*. Macmillan & Co.

On Lamarckism, Weismann, August 1889. *Essays Upon Heredity and Kindred Biological Problems*.

On epigenetics, Jablonka, Eva and Lamb, M. 2005. *Evolution in Four Dimensions: Genetic, Epigenetic and Symbolic Variation in the History of Life*. MIT Press. And Haig, D. 2007. Weismann Rules! OK? Epigenetics and the Lamarckian temptation. *Biology and Philosophy* 22:415–428.

Chapter 4: The Evolution of Genes

On the origin of life, Horgan, J. 2011. Psst! Don't tell the creationists, but scientists don't have a clue how life began. *Scientific American* 28 February 2011; and Lane, N. and Martin, W.F. 2012. The origin of membrane bioenergetics. *Cell* 151:1406–1416.

On energy and genes, Lane, Nick 2015. *The Vital Question*. Profile; and Constable, John 2014. Thermo-economics: energy, entropy and wealth. B&O Economics Research Council 44.

The calculations as to the numbers of events happening inside the human body at any one time are mine but based on information supplied by Patrick Cramer and Venki Ramakrishnan.

On selfish DNA, Dawkins, R. 1976. *The Selfish Gene*. Oxford University Press; Doolittle, W.F. and Sapienza, C. 1980. Selfish genes, the phenotype paradigm and genome evolution. *Nature* 284:601–603; and Crick, F.H.C. and Orgel, L. 1980. Selfish DNA: the ultimate parasite. *Nature* 284:604–607.

On 'junk DNA', Brosius, J. and Gould, S.J. 1992. On 'genomenclature': A comprehensive (and respectful) taxonomy for pseudogenes and other 'junk DNA'. *PNAS* 89:10706–10710. And Rains, C. 2012. No more junk DNA. *Science* 337:1581.

On defence of junk DNA, Graur, D., Zheng, Y., Price, N., Azevedo, R.B., Zufall, R.A., Elhaik, E. 2013. On the immortality of television sets: 'function' in the human genome according to the evolution-free gospel of ENCODE. *Genome Biol. Evol.* 5(3):578–590. Also Palazzo, Alexander F. and Gregory, T. Ryan 2014. The case for junk DNA. *PLOS Genetics* 10.

On the Red Queen effect, Ridley, M. 1993. *The Red Queen*. Viking.

Chapter 5: The Evolution of Culture

On embryology, Dawkins, R. 2009. *The Greatest Show on Earth*. Bantam.

On emergent order in nature, Johnson, Steven 2001. *Emergence*. Penguin.

On cultural evolution, Richerson, Peter J. and Boyd, Robert 2006. *Not by Genes Alone: How Culture Transformed Human Evolution*. University of Chicago Press; Henrich, Joe, Boyd, Robert and Richerson, Peter 2008. Five misunderstandings about cultural evolution. *Human Nature* 19:119–137; Richerson, Peter and Christiansen, Morten (eds) 2013. *Cultural Evolution: Society, Technology, Language and Religion*. MIT Press. Distin, Kate 2010. *Cultural Evolution*. Cambridge University Press.

On language, Darwin, C.R. 1871. *The Descent of Man*. Macmillan; Pagel, M. 2012. *Wired for Culture: Origins of the Human Social Mind*. Norton. Also Nettle, Daniel 1998. Explaining global patterns of language diversity. *Journal of Anthropological Archaeology* 17:354–374.

On the gradual nature of the human revolution in Africa, McBrearty, S. and Brooks, A.S. 2000. The revolution that wasn't: a new interpretation of the origin of modern human behavior. *Journal of Human Evolution* 39:453–563. Svante Pääbo's quote is from Pääbo, S. 2014. *Neanderthal Man: In Search of Lost Genomes*. Basic Books.

On cultural change driving genetic change during the human revolution, Fisher, S.E. and Ridley, M.W. 2013. Culture, genes and the human revolution. *Science* 340:929–930.

On the sexual appetite of Maurice de Saxe, see Thomas R. Philips's introduction to 'Reveries on the art of war' by Maurice de Saxe.

On human polygamy and the spread of monogamous marriage, Tucker, W. 2014. *Marriage and Civilization*. Regnery. And Henrich, J., Boyd, R. and Richerson, P. 2012. The puzzle of monogamous marriage. *Phil. Trans. Roy. Soc.* B 1589:657–669.

On cities, the lectures of Stephen Davies of the Institute of Economic Affairs, John Kay's article on 'New York's wonder shows planners' limits' in the *Financial Times* 27 March 2013; Glaeser, Edward 2011. *Triumph of the City. How Our Greatest Invention Makes Us Richer, Smarter, Greener, Healthier and Happier*. Macmillan; Geoffrey West's 2011 TED Global talk: The surprising math of cities and corporations. And Hollis, Leo 2013. *Cities are Good for You*. Bloomsbury.

On the slow pace of governmental evolution, Runciman, W.G. 2014. *Very Different, But Much the Same*. Oxford University Press.

Chapter 6: The Evolution of the Economy

On economic growth in the twenty-first century, Long-term growth scenarios. OECD Economics Department Working Papers. OECD 2012.

On the great enrichment, McCloskey, D. 2014. Equality lacks relevance if the poor are growing richer. *Financial Times* 11 August 2014. Also Phelps, Edmund 2013. *Mass Flourishing*. Princeton University Press.

On institutions, Acemoglu, D. and Robinson, J. 2011. *Why Nations Fail*. Crown Business.

On the market, Smith, Adam 1776. *The Wealth of Nations*.

William Easterly's quote is from Easterly, William 2013. *The Tyranny of Experts*. Basic Books.

On Swedish economic performance, Sanandaji, N. 2012. The Surprising ingredients of Swedish success: free markets and social cohesion. Institute of Economic Affairs.

On extravagance and conspicuous consumption, Miller, Geoffrey 2012. Sex, mutations and marketing: how the Cambrian Explosion set the stage for runaway consumerism. EMBO Reports 13:880–884. And Miller, Geoffrey 2009. *Spent: Sex, Evolution and Consumer Behavior*. Viking.

On feeding Paris, Bastiat, Frédéric 1850. *Economic Harmonies*.

On Schumpeter, McCraw, Thomas K. 2007. *Prophet of Innovation*. The Belknap Press of Harvard University Press.

McCloskey's second volume on bourgeois virtues is McCloskey, D. 2010. *Bourgeois Dignity: Why Economics Can't Explain the Modern World*. University of Chicago Press.

On economics as an evolutionary system, Hanauer, N. and Beinhocker, E. 2014. Capitalism redefined. *Democracy: A Journal of Ideas*. Winter 2014; and Beinhocker, E. 2006. *The Origin of Wealth: Evolution, Complexity, and the Radical Remaking of Economics*. Random House.

Ecological equilibrium is discussed in Marris, E. 2013. *The Rambunctious Garden: Saving Nature in a Post-Wild World*. Bloomsbury. And Botkin, Daniel 2012. *The Moon in the Nautilus Shell*. Oxford University Press. Also: Botkin, Daniel 2013. Is there a balance of nature? Danielbotkin.com 23 May 2013.

On the great enrichment, McCloskey, D. 2014. The Great Enrichment Came and Comes from Ethics and Rhetoric. Lecture, New Delhi, reprinted at deirdremccloskey.org. Also Baumol, William J., Litan, Robert E. and Schramm, Carl J. 2004. *Good Capitalism, Bad Capitalism*. Yale University Press.

On increasing returns and the search for an explanation of innovation, Warsh, David 2006. *Knowledge and the Wealth of Nations: A Story of Economic Discovery*. Norton.

Larry Summers is quoted in Easterly, William 2013. *The Tyranny of Experts*. Basic Books.

On the exchange of ideas, Ridley, Matt 2010. *The Rational Optimist*. HarperCollins.

On economic creationism, Boudreaux, Don 2013. If They Don't Get This Point, Much of What We Say Sounds Like Gibberish to Them. Blog post 5 October 2013, cafehayek.com. See also Boudreaux, Donald 2012. *Hypocrites & Half-Wits*. Free To Choose Network.

On consumers as bosses, Mises, L. von 1944. *Bureaucracy*. Available at mises.org.

Figures on healthcare and family budgets come from Conover, C.J. 2011. The Family Healthcare Budget Squeeze. *The American* November 2011. American.com.

On friendly societies, Green, D. 1985. *Working Class Patients and the Medical Establishment*. Maurice Temple Smith. And Frisby, Dominic 2013. *Life After the State*. Unbound.

Chapter 7: The Evolution of Technology

On the history of the electric light, Friedel, R. 1986. *Edison's Electric Light*. Rutgers University Press.

On simultaneous invention, Wagner, A. 2014. *Arrival of the Fittest*. Current Books; Kelly, Kevin 2010. *What Technology Wants*. Penguin (Viking); and Armstrong, Sue 2014. *The Gene that Cracked the Cancer Code*. Bloomsbury Sigma p53.

On the inevitability of the discovery of the double helix, Ridley, Matt 2006. *Francis Crick*. HarperCollins. On the four-factor formula, Spencer Weart cited in Kelly, Kevin 2010. *What Technology Wants*. Penguin (Viking).

On Moore's Law used to predict Pixar's moment, Smith, Alvy Ray 2013. How Pixar used Moore's Law to predict the future. *Wired* 17 April 2013. On Moore's Law and its cousins, Ridley, Matt 2012. Why can't things get better faster (or slower)?. *Wall Street Journal* 19 October

2012. On Moore's Law extended, Kurzweil, Ray 2006. *The Singularity is Near*. Penguin.

On evolution in technology, Arthur, W. Brian 2009. *The Nature of Technology*. Free Press; Johnson, Steven 2010. *Where Good Ideas Come From*. Penguin (Riverhead Books); Harford, Tim 2011. *Adapt*. Little, Brown; and Ridley, Matt 2010. *The Rational Optimist*. HarperCollins. George Basalla's earlier book is Basalla, George 1988. *The Evolution of Technology*. Cambridge University Press.

Alain's quip about boats is cited in Dennett, Daniel C. 2013. *Intuition Pumps and Other Tools for Thinking*. W.W. Norton & Co.

On innovation in business, Drucker, P. 1954. *The Practice of Management*. Harper Business. And Brokaw, L. 2014. How Procter & Gamble Uses External Ideas For Internal Innovation. *MIT Sloan Management Review* 16 June 2014.

On intellectual property, Tabarrok, A. 2011. *Launching the Innovation Renaissance*. TED Books.

On knowledge, Hayek, F.A. 1945. The uses of knowledge in society. *American Economic Review* 4:519–530. And Hayek, Friedrich A. *The Road to Serfdom* (Condensed Version). Reader's Digest.

On the relationship between science and technology, Kealey, Terence 2013. The Case Against Public Science. Cato-unbound.org 5 August 2013. Also Kealey, T. and Ricketts, M. 2014. Modelling science as a contribution good. *Research Policy* 43:1014–1024. Also Pielke, R. Jr 2013. Faith-based science policy. Essay at rogerpielkejr.blogspot.co.uk February 2013.

On fracking, Jenkins, Jesse, Shellenberger, Michael, Nordhaus, Ted and Trembarth, Alex 2010. US government role in shale gas fracking history: an overview and response to our critics. Breakthrough.org website, accessed 1 October 2014, and Chris Wright, personal communication.

Chapter 8: The Evolution of the Mind

Spinoza's quote about the 'thinking substance' is from the Scholium to Prop 7 of Part 2, E. Curley (trans.) 1996. Spinoza, *Ethics*. Penguin. His rolling-stone analogy and drunken man story come from Letter 62 (1674) in his *Correspondence*.

On Spinoza, Damasio, Anthony 2003. *Looking for Spinoza*. Houghton Mifflin.

On materialism and mind, Gazzaniga, Michael S. 2011. *Who's in*

Charge?. HarperCollins. Also Humphrey, Nicholas 2011. *Soul Dust: The Magic of Consciousness*. Quercus. Crick, Francis 1994. *The Astonishing Hypothesis: The Scientific Search for the Soul*. Scribner.

On experiments finding delays between action and thought, Soon, C.S., Brass, M., Heinze, H.-J., Haynes, J.D. 2008. Unconscious determinants of free decisions in the human brain. *Nature Neuroscience* 11:543–545.

On the Libet experiments, Harris, Sam 2012. *Free Will*. Free Press.

On responsibility, Cashmore, A.R. 2010. The Lucretian swerve: The biological basis of human behavior and the criminal justice system. *PNAS* 107:4499–4504.

Daniel Dennett's response to Sam Harris is Dennett, D. 2014. Reflections on free will. Review published at naturalism.org and also reprinted at samharris.org.

Robert Sapolsky is quoted in Satel, S. 2013. Distinguishing brain from mind. *The Atlantic* 13 May 2013.

On the tumour-induced paedophilia, Harris, Sam 2012. *Free Will*. Free Press.

Also Burns, J.M. and Swerdlow, R.H. 2003. Right orbitofrontal tumor with pedophilia symptom and constructional apraxia sign. *Archives of Neurology* 60:437–440.

On free will, Dennett, Daniel C. 2003. *Freedom Evolves*. Penguin.

Chapter 9: The Evolution of Personality

Judith Rich Harris's two books on nature and nurture are Harris, Judith Rich 1998. *The Nurture Assumption*. Bloomsbury; and Harris, Judith Rich 2006. *No Two Alike*. W.W. Norton.

On nature–nurture, Pinker, S. 2002. *The Blank Slate: The Modern Denial of Human Nature*. Allen Lane. And Ridley, Matt 2003. *Nature via Nurture*. HarperCollins.

On genes that influence behaviour, Weiner, J. 1999. *Time, Love, Memory: A Great Biologist and his Quest for Human Behavior*. Knopf.

On 'not in our genes', Lewontin, R., Rose, S. and Kamin, L. 1984. *Not in Our Genes: Ideology and Human Behavior*. Pantheon.

On genes and intelligence, Plomin, R., Haworth, C.M.A., Meaburn, E.L., Price, T.S. and Davis, O.S.P. 2013. Common DNA markers can account for more than half of the genetic influence on cognitive abilities. *Psychological Science* 24:562–568. Plomin, Robert, Shakeshaft, Nicholas G., McMillan, Andrew and Trzaskowski, Maciej 2014.

Nature, nurture, and expertise. *Intelligence* 45:46–59. Also Plomin, R., DeFries, J.C., Knopik, V.S. and Neiderhiser, J.M. 2013. *Behavioral Genetics* (6th edition). Worth Publishers.

On intelligence heritability increasing with age, Briley, D.A. and Tucker-Drob, E.M. 2013. Explaining the increasing heritability of cognitive ability over development: A meta-analysis of longitudinal twin and adoption studies. *Psychological Science* 24:1704–1713; and Briley, D.A. and Tucker-Drob, E.M. 2014. Genetic and environmental continuity in personality development: A meta-analysis. *Psychological Bulletin* 140:1303–1331.

On regression to the mean, Clark, Gregory 2014. *The Son Also Rises*. Princeton University Press.

On monkeys and toys, Hines, M. and Alexander, G.M. 2008. Monkeys, girls, boys and toys: A confirmation letter regarding 'Sex differences in toy preferences: Striking parallels between monkeys and humans'. *Horm. Behav.* 54:478–479.

On universal similarity of homicide patterns, Daly, M. and Wilson, M. 1988. *Homicide*. Aldine.

On age preferences of men and women, Buunk, P.P., Dujkstra, P., Kenrick, D.T. and Warntjes, A. 2001. Age preferences for mates as related to gender, own age, and involvement level. *Evolution and Human Behavior* 22:241–250.

Chapter 10: The Evolution of Education

On Prussian schools, Rothbard, M. 1973. *For a New Liberty*. Collier Macmillan.

On literacy rates, Clark, G. 2007. *A Farewell to Alms: A Brief Economic History of the World*. Princeton University Press.

On Edwin West, West, Edwin G. 1970. Forster and after: 100 years of state education. *Economic Age* 2.

On low-cost private education, Tooley, James 2009. *The Beautiful Tree: A Personal Journey into How the World's Poorest People are Educating Themselves*. Cato Institute. And Tooley, James 2012. *From Village School to Global Brand*. Profile Books.

On the public purpose of public education and on the starfish and spider models, Pritchett, Lant 2013. *The Rebirth of Education: Schooling Ain't Learning*. Brookings Institution Press.

On markets in education, Coulson, A. 2008. Monopolies vs. markets in education: a global review of the evidence. Cato Institute, Policy Paper no 620.

Other sources: Frisby, D. 2013. *Life After the State*. Unbound. Stephen Davies, Institute of Economic Affairs lectures.

Einstein quote from Einstein, A. 1991. *Autobiographical Notes*. Open Court.

Albert Shanker quote from Kahlenberg, R.D. 2007. *Tough Liberal: Albert Shanker and the Battles Over Schools, Unions, Race and Democracy*. Columbia University Press.

On Swedish schools, Stanfield, James B. 2012. *The Profit Motive in Education: Continuing the Revolution*. Institute of Economic Affairs.

On MOOCs, Brynjolfsson, E. and McAfee, A. 2014. *The Second Machine Age*. Norton.

On Minerva College, Wood, Graeme. The future of college?. *The Atlantic* September 2014.

Sugata Mitra's TED talks are available at TED.com. His short book is *Beyond the Hole in the Wall: Discover the Power of Self-Organized Learning*. TED Books 2012.

On environmental indoctrination, Montford, A. and Shade, J. 2014. Climate Control: brainwashing in schools. Global Warming Policy Foundation.

On Montessori schools, Sims, P. 2011. The Montessori Mafia. *Wall Street Journal* 5 April 2011.

Alison Wolf's studies are described in Wolf, A. 2002. *Does Education Matter?*. Penguin; and Wolf, Alison 2004. The education myth. Project-syndicate.org. Also Wolf, A. 2011. Review of Vocational Education: The Wolf Report. UK Government.

Chapter 11: The Evolution of Population

On the connection between nineteenth-century Malthusian ideas and twentieth-century eugenics and population controls, Zubrin, Robert 2012. *Merchants of Despair*. Encounter Books (New Atlantis Books); Desrochers, P. and Hoffbauer, C. 2009. The Post War Intellectual Roots of the Population Bomb. Fairfield Osborn's 'Our Plundered Planet' and William Vogt's 'Road to Survival'. *Retrospect. The Electronic Journal of Sustainable Development* 1:37–51.

On the Irish famine, Pearce, F. 2010. *The Coming Population Crash*. Beacon.

On Darwin's eugenics brush, Darwin, C.R. 1871. *The Descent of Man*. Macmillan. On Galton's eugenics, Pearson, Karl 1914. *Galton's Life and Letters*. Cambridge University Press.

Ernst Haeckel's Altenburg lecture is 'Monism as connecting science and faith' (1892).

On Malthusian and eugenic enthusiasms before the First World War, Macmillan, Margaret 2013. *The War that Ended Peace*. Profile.

On liberal fascism, Goldberg, Jonah 2007. *Liberal Fascism*. Doubleday.

On Madison Grant's role, Wade, N. 2014. *A Troublesome Inheritance*. Penguin.

On the environmental enthusiasm of the Nazis, Durkin, M. 2013. Nazi Greens – an inconvenient history. At Martindurkin.com.

On the post-war population movement, Mosher, S.W. 2003. The Malthusian Delusion and the Origins of Population Control. *PRI Review* 13.

On 1960s population books, Paddock, W. and Paddock, P. 1967. *Famine 1975!*. Little, Brown. And Ehrlich, P. 1968. *The Population Bomb*. Ballantine. Also Ehrlich, P., Ehrlich, A. and Holdren, J. 1978. *Ecoscience*. Freeman.

On the demographic transition, Hanson, Earl Parker 1949. *New Worlds Emerging*. Duell, Sloan & Pearce. And Castro, J. de. 1952. *The Geopolitics of Hunger*. Monthly Review Press.

On resources, Simon, Julian 1995. Earth Day: Spiritually uplifting, intellectually debased. Essay available at juliansimon.org.

On the Club of Rome, Delingpole, J. 2012. *Watermelons: How Environmentalists are Killing the Planet, Destroying the Economy and Stealing Your Children's Future*. Biteback. The Club's 1974 manifesto is at 'Mankind at the Turning Point'. Also Goldsmith, E. 1972. *A Blueprint for Survival*. Penguin.

On China's one-child policy, Greenhalgh, S. 2005. Missile Science, Population Science: The Origins of China's One-Child Policy. *China Quarterly* 182:253–276; Greenhalgh, S. 2008. *Just One Child: Science and Policy in Deng's China*. University of California Press. Also: Ted Turner urges global one-child policy to save planet. *Globe and Mail* 5 December 2010.

The video of Jacob Bronowski's remarks at the end of *The Ascent of Man* is available on the internet.

Chapter 12: The Evolution of Leadership

On Montesquieu and great men, Macfarlane, Alan 2000. *The Riddle of the Modern World*. Palgrave. Mingardi, Alberto 2011. *Herbert Spencer*. Bloomsbury Academic.

On Churchill, Johnson, B. 2014. *The Churchill Factor: How One Man Made History*. Hodder & Stoughton.

On Chinese reform: The secret document that transformed China. National Public Radio report on Chinese land reform 14 May 2014.

On the American presidency, Bacevich, A. 2013. The Iran deal just shows how badly Obama has failed. *Spectator* 30 November 2013.

On the impact of Gutenberg, Johnson, S. 2014. *How We Got to Now*. Particular Books.

On mosquitoes and wars, Mann, Charles C. 2011. *1493*. Granta Books. And McNeill, J.R. 2010. Malarial mosquitoes helped defeat British in battle that ended Revolutionary War. *Washington Post* 18 October 2010.

On imperial chief executives, Johnson, Steven 2012. *Future Perfect*. Penguin. And Hamel, G. 2011. First Let's Fire All the Managers. *Harvard Business Review* December 2011.

On Morning Star Tomatoes, I, Tomato: Morning Star's Radical Approach to Management. Available on YouTube. And Green, P. 2010. The Colleague Letter of Understanding: Replacing Jobs with Commitments. Managementexchange.com.

On self-management, Wartzman, R. 2012. If Self-Management is Such a Great Idea, Why Aren't More Companies Doing It?. *Forbes* 25 September 2012.

On economic development, Rodrik, D. 2013.The Past, Present, and Future of Economic Growth. Global Citizen Foundation. And Easterly, William 2013. *The Tyranny of Experts*. Basic Books. Also McCloskey, D. 2012. Factual Free-Market Fairness. Bleedingheartlibertarians.com. And Lal, Deepak 2013. *Poverty and Progress*. Cato Institute. And: Villagers losing their land to Malawi's sugar growers. BBC News 16 December 2014.

Chapter 13: The Evolution of Government

On the wild west, Anderson, Terry and Hill, P.J. 2004. *The Not So Wild, Wild West*. Stanford Economics and Finance.

On prisons, Skarbek, D. 2014. *The Social Order of the Underworld: How Prison Gangs Govern the American Penal System*. Oxford University Press.

On governments as organised crime, Williamson, Kevin D. 2013. *The End is Near and it's Going to be Awesome*. HarperCollins; Nock, A.J. 1939. The criminality of the state. *The American Mercury* March 1939; and Morris, Ian 2014. *War: What is it Good For?*. Farrar, Straus & Giroux. Also Robert Higgs, Some basics of state domination and public submission. Blog.independent.org 27 April 2104.

On Ferguson, Missouri, Paul, Rand. We must demilitarize the police. *Time* 14 August 2014. Balko, Radley 2013. *Rise of the Warrior Cop*. PublicAffairs.

On Lao Tzu, Blacksburg, A. 2013. Taoism and Libertarianism – From Lao Tzu to Murray Rothbard. Thehumanecondition.com.

Lord Acton's letter to Mary Gladstone (24 April 1881), published in *Letters of Lord Acton to Mary Gladstone* (1913) p. 73. Michael Cloud quoted in Frisby, Dominic 2013. *Life After the State*. Unbound.

On the Levellers, see 'An arrow against all tyrants' by Richard Overton, 12 October 1646, available at constitution.org. And Hannan, Daniel 2013. *How We Invented Freedom and Why it Matters*. Head of Zeus Ltd.

On eighteenth-century liberalism, the lectures of Stephen Davies, online at IEA.com are especially good.

On the history of government, Micklethwait, John and Wooldridge, Adrian 2014. *The Fourth Revolution*. Allen Lane.

On the politics of Adam Smith, see Rothschild, Emma 2001. *Economic Sentiments: Adam Smith, Condorcet and the Enlightenment*. Harvard University Press.

On Hamilton and Jefferson, see Will, George 2014. Progressives take lessons from 'Downton Abbey'. *Washington Post* 12 February 2014.

On British liberal thinking, Martineau, Harriet 1832–1834. *Illustrations of political economy*. Also Micklethwait, John and Wooldridge, Adrian 2014. *The Fourth Revolution*. Allen Lane.

On free trade, Bernstein, William 2008. *A Splendid Exchange: How Trade Shaped the World*. Atlantic Monthly Press. Also Lampe, Markus 2009. Effects of bilateralism and the MFN clause on international trade – Evidence for the Cobden-Chevalier Network (1860–1875). dev3.cepr.org. And Trentman, Frank 2008. *Free Trade Nation*. Oxford University Press.

On the industrial counter-revolution, Lindsey, Brink 2002. *Against the Dead Hand*. John Wiley & Sons; Dicey, A. V. [1905] 2002. Lectures on the Relation between Law and Public Opinion in England during the Nineteenth Century.

On twentieth-century liberalism, Goldberg, Jonah 2007. *Liberal Fascism*. Doubleday. Brogan, Colm 1943. *Who are 'the People'?*. Hollis & Carter. Agar, Herbert 1943. *A Time for Greatness*. Eyre & Spottiswoode.

On the growth of government, Micklethwait, John and Wooldridge, Adrian 2014. *The Fourth Revolution*. Allen Lane.

Christiana Figueres, interview with Yale Environment 360. Printed in the *Guardian* 21 November 2012.

On the future evolution of politics, Carswell, Douglas 2012. *The End of Politics and the Birth of iDemocracy*. Biteback.

Chapter 14: The Evolution of Religion

On religion, O'Grady, Selina 2012. *And Man Created God*. Atlantic Books; Armstrong, Karen 1993. *A History of God*. Knopf; Wright, Robert 2009. *The Evolution of God*. Little, Brown; Baumard, N. and Boyer, P. 2013. Explaining moral religions. *Trends in Cognitive Sciences* 17:272–280; Holland, T. 2012. *In the Shadow of the Sword*. Little, Brown; Birth of a religion. Interview with Tom Holland, *New Statesman* 3 April 2012.

On crop circles, the television programme referred to is *Equinox: The Strange Case of Crop Circles* (Channel 4, UK 1991); the book that thinks the CIA and the Vatican are out to debunk them is Silva, Freddy 2013. *Secrets in the Fields*. Invisible Temple.

On the yearning to believe, Steiner, George 1997. Nostalgia for the Absolute (CBC Massey Lecture). House of Anansi.

On pigeons, Skinner, B.F. 1947 'Superstition' in the Pigeon. *Journal of Experimental Psychology* 38:168–172.

On pseudoscience, Popper, K. 1963. *Conjectures and Refutations*. Routledge & Keegan Paul; Shermer, Michael 2012. *The Believing Brain: From Ghosts and Gods to Politics and Conspiracies – How We Construct Beliefs and Reinforce Them as Truths*. St Martin's Griffin.

On vitalism, Crick, Francis 1966. *Of Molecules and Men*. University of Washington Press.

On biodynamic farming, Chalker-Scott, Linda 2004. The myth of biodynamic agriculture. Puyallup.wsu.edu.

On climate, Curry, Judith 2013. CO_2 'control knob' theory. judithcurry. com 20 September 2013. On CO_2 and ice ages, Petit, J.R. et al. 1999. Climate and atmospheric history of the past 420,000 years from the Vostok ice core, Antarctica. *Nature* 399:429–436; and Eschenbach, Willis 2012. Shakun Redux: Master tricksed us! I told you he was tricksy! Wattsupwiththat.com 7 April 2012. Goklany, I. 2011. Could biofuel policies increase death and disease in developing countries?. *Journal of American Physicians and Surgeons* 16:9–13. Bell, Larry. Climate Change as Religion: The Gospel According to Gore. *Forbes* 26 April 2011. Lilley, Peter 2013. Global Warming as a 21st Century Religion. *Huffington Post* 21 August 2013. Bruckner, Pascal 2013. Against environmental panic. *Chronicle Review* 27 June 2013. Bruckner, Pascal 2013. *The Fanaticism of the Apocalypse: Save the Earth, Punish Human Beings*. Polity Press. Lawson, Nigel 2014. *The Trouble With Climate Change*. Global Warming Policy Foundation.

On floods, O'Neill, Brendan 2014. The eco-hysteria of blaming mankind for the floods. *Spiked* 20 February 2014.

On weather, Pfister, Christian, Brazdil, Rudolf and Glaser, Rudiger 1999. *Climatic Variability in Sixteenth-Century Europe and its Social Dimension: A Synthesis*. Springer.

On deaths caused by weather, Goklany, I. 2009. Deaths and Death Rates from Extreme Weather Events: 1900–2008. *Journal of American Physicians and Surgeons* 14:102–109.

Chapter 15: The Evolution of Money

On Birmingham tokens, Selgin, George 2008. *Good Money*. University of Michigan Press.

On central banks, Ahamed, Liaquat 2009. *Lords of Finance*. Windmill Books. Norberg, Johan 2009. *Financial Fiasco*. Cato Institute. And Selgin, George 2014. William Jennings Bryan and the Founding of the Fed. Freebanking.org 20 April 2014. Also Taleb, N.N. 2012. *Antifragile*. Random House.

On dollarisation, Allister Heath. The Scottish nationalists aren't credible on keeping sterling. *City AM* 14 February 2014.

On regulation, Gilder, George 2013. *Knowledge and Power*. Regnery.

On Fannie and Freddie, Stockman, David A. 2013. *The Great Deformation*. PublicAffairs; Woods, Thomas E. Jr 2009. *Meltdown*. Regnery; Kurtz, Stanley 2010. *Radical in Chief*. Threshold Editions;

Krugman, Paul 2008. Fannie, Freddie and you. *New York Times* 14 July 2008.

On the financial crisis, Norberg, Johan 2009. *Financial Fiasco*. Cato Institute; Atlas, John 2010. *Seeds of Change*. Vanderbilt University Press; Allison, John A. 2013. *The Financial Crisis and the Free Market Cure*. McGraw-Hill. Friedman, Jeffrey (ed.) 2010. *What Caused the Financial Crisis*. University of Pennsylvania Press. Wallison, Peter 2011. The true story of the financial crisis. *American Spectator* May 2011. And Booth, Philip (ed.) 2009. Verdict on the Crash. IEA.

On the Cantillon Effect, Frisby, Dominic 2013. *Life After the State*. Unbound.

On mobile money, Why does Kenya lead the world in mobile money?. economist.com 27 May 2013.

On the Federal Reserve, Selgin, G., Lastrapes, W.D. and White, L.H. 2010. Has the Fed been a Failure? Cato Working Paper, Cato.org. Hsieh, Chang-Tai and Romer, Christina D. 2006. Was the Federal Reserve Constrained by the Gold Standard During the Great Depression? Evidence from the 1932 Open Market Purchase Program.*Journal of Economic History* 66(1) (March):140–176. And Selgin, George 2014. William Jennings Bryan and the Founding of the Fed. Freebanking.org 20 April 2014.

Chapter 16: The Evolution of the Internet

Hayek quote from Hayek, F. 1978. *The Constitution of Liberty*. University of Chicago Press.

On East German televisions, and telephones, Kupferberg, Feiwel 2002. *The Rise and Fall of the German Democratic Republic*. Transaction Publishers.

On the Arpanet, Crovitz, Gordon 2012. Who really invented the internet?. *Wall Street Journal* 22 July 2012.

On peer-to-peer networks, Johnson, Steven 2012. *Future Perfect*. Penguin.

On the balkanisation of the web, Sparkes, Matthew 2014. The Coming Digital Anarchy. *Daily Telegraph* 9 June 2014.

On Wikipedia editing, Scott, Nigel 2014. Wikipedia: where truth dies online. *Spiked* 29 April 2014. Filipachi, Amanda 2013. Sexism on Wikipedia is Not the Work of 'a Single Misguided Editor'. *The Atlantic* 13 April 2013. Solomon, Lawrence 2009. Wikipedia's climate doctor. Nationalpost.com (no date). Also: Global warming propagandist slapped down by Wikipedia. sppiblog.org.

On permissionless innovation, Cerf, Vinton 2012. Keep the Internet Open. *New York Times* 23 May 2012. And Thierer, A. 2014. *Permissionless Innovation: The Continuing Case for Comprehensive Technological Freedom*. Mercatus Center, George Mason University.

On the ITU, Blue, Violet 2013. FCC to Congress: U.N.'s ITU Internet plans 'must be stopped'. zdnet.com 5 February 2013.

On net censorship, MacKinnon, Rebecca 2012. *Consent of the Networked*. Basic Books.

On blockchains, Frisby, Dominic 2014. *Bitcoin: The Future of Money?*. Unbound.

On Nick Szabo's 'shelling out', nakamotoinstitute.org/shelling-out/.

On Ethereum's white paper, A Next-Generation Smart Contract and Decentralized Application Platform. https://github.com/ethereum.

On private money, Dowd, K. 2014. *New Private Monies*. IEA.

On smart contracts, De Filippi, P. 2014. Ethereum: freenet or skynet?. At cyber.law.harvard.edu/events 14 April 2014.

On digital politics, Carswell, Douglas 2014. iDemocracy will change Westminster for the Better. Govknow.com 20 April 2014. And Carswell, Douglas 2012. *The End of Politics and the Birth of iDemocracy*. Biteback. Also Mair, Peter 2013. *Ruling the Void*. Verso.

Epilogue: The Evolution of the Future

On Sir David Butler's point about incremental changes having little to do with government action, interview with Sir Andrew Dilnot on BBC Radio 4, 27 February 2015.

On unordered phenomena, Lindsey, Brink 2002. *Against the Dead Hand*. John Wiley & Sons.

INDEX

Abd al-Malik 262–3
Acemoglu, Daron 97–8
Act of Union (UK, 1707) 281, 283
Acton, John Dalberg-Acton, Lord 219, 241–2
Adams, Douglas 20
Adams, John 120–1
Adler, Alfred 269
Afghanistan 32, 258
Africa 82–5, 87, 134, 158, 183, 194, 197, 206, 229, 231, 233
Africa Governance Initiative 232
Agar, Herbert 252
AIG 287, 294
Akbar 87
Aktion T4 programme (1939) 203
Al Khwarizmi 119
Al Qaeda 3
Alaska 80, 81, 82
Alexander the Great 262
Allison, John, *The Financial Crisis and the Free Market Cure* 293
Altavista 120
Amazon (company) 188
American Eugenics Society 204
American Federation of Teachers 180
American Museum of Natural History 200, 201

American Revolution 220–2, 243, 250, 282n
American University 139
Ammon, Otto 198
Anderson, Terry 235–6
Anti-Corn Law League 245
Antonopoulos, Andreas 313–14
Apollonius of Tyana 257–8
Apple Computer 223, 319
Aquinas, Thomas 39, 51
Arabia 86, 260
Archie (search engine) 120
Arendt, Hannah 253
Argentina 190
Aristotle 8, 11
Arnhart, Larry 27
Arpanet 300–1
Arrow, Kenneth 137, 138
Arthur, Brian, *The Nature of Technology: What it is and How it Evolves* 126
Asia 82, 86, 229, 303
Assange, Julian 303
Association of Community Organizers for Reform Now (ACORN) 290–2
Athens 101
Attila the Hun 87
Augustus, Emperor 239, 257

Auschwitz-Birkenau 193, 198, 214
Australia 34, 82, 123, 244
Austria 32, 247
Ayr Bank 282
Aztecs 259

Back, Adam 307
Bacon, Francis 15, 134
Bagehot, Walter 297; *Lombard Street* 285
Baldwin effect 57
Balko, Radley 241
Balliol College, Oxford 22
Bank of England 278, 281, 282, 283, 284, 285, 295
Bank of Scotland 281
Banque Royale 285–6
Baran, Paul 300, 301
Barlow, John 302
Basalla, George, *The Evolution of Technology* 128
Bastiat, Frédéric 102; *Economic Harmonies* 102
Bath 91
Baumard, Nicolas 259, 260
Bayle, Pierre, *Thoughts on the Comet of 1680* 16
Beagle (ship) 38
Behe, Professor Michael 50–1; *Darwin's Black Box* 51
Behringer, Wolfgang 276
Beijing 193
Beinhocker, Eric 107
Belgium 32
Bell, Alexander Graham 119, 200
Bell, Andrew 186; *An Experiment in Education, made at the Male Asylum at Madras; suggesting a System by which a School or Family may teach itself, under the Superintendence of the Master or Parent* 186
Bellamy, Edward 249; *Looking Back* 249–50

Belloc, Hilaire 95
Benjamin, Park, *The Age of Electricity* 120
Benn, Sir Ernest 253
Bentham, Jeremy 35
Berkeley, Bishop George 12
Berlin, Isaiah 253
Berlin, Steven 299
Bernick, Evan 241
Beverley, Robert Mackenzie 43
Bezos, Jeff 222, 223
Big Pharma 133
Birmingham 91
Bismark, Otto von 247
bitcoin 298, 308–9, 310–12
Blackbird, reconnaissance plane 130
Blair, Tony 232
blockchain 313–14
Blockchain.info 313
A Blueprint for Survival (Club of Rome) 211–12
Blunt, John 285
Bodanis, David, *Passionate Minds* 20
Boston Tea Party 282n
Botkin, Daniel 108
Botticelli, Sandro, *Venus* 12
Boudreaux, Don 35, 111, 113
Boulton, Matthew 278, 280
Bower, Doug 265–6
Boyd, Rob 78, 89
Boyer, Pascal 259, 260
Boyle, Robert 12
Boyle's Law 120
Bracciolini, Gian Francesco Poggio 12
Bracken, Mike 255
Brahmagupta 119
Brasilia 92
Brazil 18, 125
Brenner, Sydney 70
Bridge International Academies group 184

Bright, John 246
Brin, Sergey 119
Bristol 91
British Eugenics Society 204
British Linen Bank 281
British Medical Association 115
Brogan, Colm, *Who are 'the People'?* 252–3
Bronowski, Jacob, *The Ascent of Man* 214
Bronze Age 91
Brosius, Jürgen 69
Brown, Gordon 315
Bruckner, Pascal 274
Bruno, Giordano 12
Bryan, William Jennings 49, 284–5
Buccleuch, Henry Scott, 3rd Duke of 22
Buddha 257
Buddhism 259, 260
Bulwer-Lytton, Robert 196
Bureau of Economic Analysis 139
Burford, Oxfordshire 242
Bush, George W. 50, 290, 291, 293
Butler, Sir David 318
Butler, Samuel 127
Byron, George Gordon, 6th Lord 16, 248

California 201–2, 223, 225–8, 233–4, 236, 238, 302
Callaghan, James 295
Calvinists 142
Canada 32, 34, 170, 177, 284
Canberra 92
Cantillon, Richard 294
Cantillon Effect 294–5
Cape Horn 82
Carlyle, Thomas 216, 217
Carroll, Lewis, *Through the Looking Glass* 73
Carswell, Douglas 242, 243, 295, 314–16; *The End of Politics and the Birth of iDemocracy* 255

Cashmore, Anthony 148, 151
Castlereagh, Robert Stewart, Viscount 245
Castro, Fidel 252
Castro, José de, *The Geopolitics of Hunger* 209
Catholic Church 3
Catholics 142
Catmull, Ed 124
Cato Institute 179
Central African Republic 32
Cerf, Vint 301, 305
CERN, Switzerland 138
Chamberlain, Joseph 248
Chamberlain, Neville 248
chaos theory 18
Charles II 88
Charles V 101
Charleton, Walter 13
Chartier, Emile ('Alain') 128
Chartists 245
Chase Manhattan 290
Cheltenham 91
Chen Yun 212
Chengdu 212
Chesterton, G.K. 268
Chiang Kai-shek 230–1
Chicago 171
China 87, 95, 105–6, 125, 181, 210, 213, 214, 217–19, 230–1, 259, 278, 288, 303, 305
Chinese Revolution 318
Chorley, Dave 265–6
Christianity 88, 258, 259, 260, 261, 263, 269, 274
Church of Jesus Christ of Latter Day Saints (Mormons) 89, 263
Churchill, Winston 197, 199, 217, 253, 295
Ciba Foundation 205
Cicero, Marcus Tullius 9
Cincinnati University 130
Clarendon, George Villiers, 4th Earl 195

Clark, Gregory 167
Clay, General Lucius 253
Cleese, John 42
Clemmer, Donald 237
Cleveland, Grover 284
climate change 271–6
Clinton, Bill 211, 291, 292, 293
Cloud, Michael 242
Club of Rome 211
Clydesdale Bank 281
Cobden, Richard 245–7, 249, 253
Cobden-Chevalier Treaty 247
Coca-Cola 112
Cold War 254, 302, 316
Colombia 232
Combine 115
Commercial Bank of Scotland 281
Communism 251, 252, 274, 303
The Communist Manifesto (Marx & Engels) 248
Communist Party 3
Community Reinvestment Act (US, 1977) 290
Company of Moneyers 279
Condliffe, John Bell 231
Condorcet, Marquis de 243, 244
Confucianism 259
Confucius 257
'Connect and Develop' project 130
Conservative Foundation 204
Constable, John 63
Cooper's Law 124
Copernicus, Nikolaus 12, 45
Corn Laws 246
Cornwallis, Charles 220–2
Cosmides, Leda 43
Coughlin, Father Charles 251
Coulson, Andrew 179
Cowperthwaite, Sir John 233–4
Creationists 50, 60
Crichton, Michael 273
Crick, Francis 59–60, 67, 121, 141, 145–6, 148, 271
Crimean War 245

Cromwell, Oliver 88, 242, 267
Cuba 303
culture: and cities 76, 91–3; and human revolution 82–5; and institutions 94–5; and language 79–82; and marriage 85–90; natural phenomena 76–8
Curry, Judith 272
cypherpunks 306–7
Cyprus 311

Dai, Wei 307
Dalai Lama 211
d'Alembert, Jean-Baptiste le Rond 142, 215
Daly, Martin 171
Damasio, Anthony 143
Dark Ages 88
Darrow, Clarence 49
Darwin, Charles 4, 9, 20, 37–8, 41, 42–5, 49–50, 52–4, 60, 110, 120, 196, 198, 319; The Descent of Man 79, 196–7
Darwin, Sir Charles Galton, The Next Million Years 204–5
Darwin, Erasmus 16; The Temple of Nature 16
Darwin, Leonard 199, 200
Darwin, Robert 245
Darwinism 52, 53–8, 73, 89, 110, 126, 128, 255
Davenport, Charles 201
Davies, Donald 301
Davies, Stephen 175–6, 189, 243
Dawkins, Richard 11, 44, 46, 53, 58, 65–8, 71, 268, 307; The Greatest Show on Earth 76; The Selfish Gene 66–7
De Filippi, Primavera 313
Dead Sea scrolls 261
Defoe, Daniel 94
Delgado, Pat 265–6
Delhi 185
Democratic Republic of Congo 32

Democritus 8, 14
Deng Xiaoping 212, 217–19
Denisovans 82
Denmark 32
Dennett, Daniel 7, 39, 40, 41, 53, 61, 65, 149–50, 151, 256–7; *Darwin's Dangerous Idea* 42–3; *Freedom Evolves* 153
Dennis, Daniel 87
Descartes, René 12, 141, 145
Detroit 93
d'Holbach, Paul-Henri Thiry, baron 22; *Le Système de la Nature* 17
Dicey, A.V. 249
Diderot, Denis 20, 142, 215, 219; *Encyclopédie* 22; *The Letter on the Blind and the Deaf* 16; *Philosophical Thoughts* 16
Diggers 242
Disney 124
Disraeli, Benjamin 196
DNA 46, 56, 60, 61, 66–72, 78, 79, 145
Donne, John 15
Doolittle, Ford 67
Dover, Gabby, *Dear Mr Darwin* 73
Dover School District 50, 51
Down, Kevin 311
Draper, General William 206
Dread Pirate Roberts 311
Drucker, Peter, *The Practice of Management* 128–9
Dryden, John 11, 15
Durham University 311
Durkheim, Emile 165
Durkin, Martin 201
Dutch republic 101

Eagleman, David 151
East Asia 288
East Germany 300
East India Company 282n
East India Company College 195

Easterly, William 100, 228–30
Economist 171
economy 4–5; and aid 229–33; background 96–8; commerce and freedom 243–4; consumerism 111–14, 129; and creationism 229; and creative destruction 107; diminishing returns 104–8; exchange 111; free trade and free thinking 244–6; friendly societies 115–16; health care 115–17; human action/human design 98–100; imperfect markets 100–3, 107–8; innovation 106, 108–10; invisible hand 103–4; perfect market 107; and the state 114–15; variation and selection 110–11
Ecuador 286
Edinburgh 91, 244
Edison, Thomas 119
education: compulsory 174–5; and economic growth 188–92; and indoctrination 187–8; innovation in 179–84; and private schools 176–9; Prussian model 175–6; technology of 184–7
Education Act (1870) 178
Egypt, Egyptians 86, 87, 101, 123, 190, 207, 259
Ehrlich, Paul 205, 209; *The Population Bomb* 207–8
Einstein, Albert 121, 180, 269
Eisenhower, Dwight D. 206
El Salvador 286
Elias, Norbert 29–30, 32, 33
Ellickson, Robert 236
Ellis, Havelock 197
Emilie, Marquise du Châtelet 15, 17
Empedocles 52
ENCODE 70–1

Encyclopaedia Britannica (1911) 216

Encyclopédie 142–3, 215, 216

Engels, Friedrich 194, 248

Enlightenment 64, 143, 215, 244, 272

Environment Defense Fund 204

Epicureans 40, 52

Epicurus 8, 14

Erhard, Ludwig 253

Ericsson 101

Ethereum 308, 313

Ethiopia 232

The Eugenics Record Office, Cold Spring Harbor (New York) 200

Europe 12, 29, 34, 82, 87, 103, 232, 247, 269, 303

European Union 35, 254

European Union Court of Justice 304–5

Evelyn, John 15

evolution: adaptation theory 2; cause and effect 3; definition 1; and evolutionary phenomena 4–5; general theory of 5–6; as gradual, incremental, undirected, emergent 1–2; special theory of 5; teaching of human history 2–3; world as self-organising/self-changing 5

Excite (search engine) 120

FARC (Revolutionary Armed Forces of Colombia) 240

fascism 251, 252, 253

Federal Deposit Insurance Corporation (FDIC) 287

Federal Home Loan Mortgage Corporation ('Freddie Mac') 289–94

Federal National Mortgage Association ('Fannie Mae') 289–94

Ferguson, Adam 4, 238

Ferguson, St Louis 240–1

Feynman, Richard 272

Fibonacci (Leonardo Binacci) 119

Figueres, Christiana 254

financial crisis (2008–09) 287–94, 307, 318

Financial Stability Board 254

Finland 32

Finn, Bernard 119

Finney, Hal 307, 310

First International Congress of Eugenics 199

First World War 198, 247, 248, 250, 318

Fisher, Simon 84

Flaubert, Gustave 9, 249

Fong, H.D. 230–1

Forster, W.E. 178

Fort Worth, Texas 136

The Fourth Revolution (Wooldridge & Micklethwait) 247

France 29, 101, 122, 138, 202, 247

Francis, Pope, *Evangelii Gaudium* 32

Franco–Prussian War 247

Frank, Barney 292

Franklin, Benjamin 20

Franklin, Rosalind 121

Frederick Augustus, Elector of Saxony, King of Poland 88

Frederick the Great 243

French Revolution 17, 243

Freud, Sigmund 156, 165, 269

Friedel, Robert 119

Friedman, Jeff 294

Friedman, Milton 306

Frisby, Dominic 115, 310, 312; *Life After the State* 294–5

Fulda, monastery at 12

Furberg, Sven 121

future, evolution of 317–20

Gaia 20, 256, 264, 269

Galaxy (search engine) 120

Galileo Galilei 12, 20
Galton, Francis 197
Gandhi, Indira 206–7
Gandhi, M.K. 'Mahatma' 178
Garzik, Jeff 312
Gas Research Institute 136
Gassendi, Pierre 12, 13
Gates, Bill 222
Gaua 81
Gazzaniga, Michael 144, 147
GCHQ 303
genes: background 59–61; function
    of 65; and the genome 62–4;
    and junk or surplus DNA
    66–72; mutation 72–5; selfish
    gene 66, 68
Genghis Khan 87, 223
geology 17
George III 245
Georgia Inst. of Technology 272
German Society for Racial Hygiene
    198, 202
Germany 12, 29, 101, 122, 138,
    231, 243, 247, 251, 253, 318
Ghana 181, 229
Giaever, Ivar 273
Gilder, George 287
Gilfillan, Colum 127
Gladstone, William Ewart 246
Glaeser, Edward 92
Glasgow University 22, 25
Glass-Steagall Act 287
global warming 271–6
Glorious Revolution (England) 243
Gobi desert 92
Goddard, Robert 138
Godkin, Ed 250
Goethe, Charles 202
Goethe, Johann Wolfgang von 248
Goldberg, Jonah 252; Liberal
    Fascism 199, 251
Goldman Sachs 3
Goldsmith, Sir Edward 211
Goodenough, Oliver 36

Google 120, 130, 132, 188
Gore, Al 205, 211, 273, 274
Gosling, Raymond 121
Gottlieb, Anthony 41
Gottlieb, Richard 11
Gould, Stephen Jay 38, 53, 69
government: commerce and
    freedom 243–4; counter-
    revolution of 247–50; definition
    236; free trade and free thinking
    244–6; as God 254–5; and the
    Levellers 241–2; liberal fascism
    250–2; libertarian revival
    252–3; prison system 237–8;
    and protection rackets 238–41;
    and the wild west 235–6
Grant, Madison 202; The Passing
    of the Great Race 200–1
Graur, Dan 71, 72
Gray, Asa 44; Descent of Man 44–5
Gray, Elisha 119
Great Depression 105, 125, 318
Great Recession (2008–09) 97, 297
Greece 259
Green, David 115
Green, Paul 226
Green Revolution 208, 210
Greenblatt, Stephen 9, 11n
Greenhalgh, Susan 212; Just One
    Child 210–11
Greenspan Put 289
Gregory, Ryan 71
Gregory VII, Pope 239
Gresham's Law 279
Guardian (newspaper) 53
Gulf War 298
Gutenberg, Johannes 220

Hadiths 262
Haeckel, Ernst 197, 198
Hahnemann, Samuel 271
Haig, David 57
Hailey, Malcolm, Lord 231
Hailo 109

Haiti 207
Hamel, Gary 224
Hamilton, Alexander 244
Hannan, Daniel 35, 242, 315
Hannauer, Nick 107
Hansen, Alvin 105
Hanson, Earl Parker, *New Worlds Emerging* 209
Harford, Tim, *Adapt: Why Success Always Starts With Failure* 127, 255
Harriman, E.H. 200
Harris, Judith Rich 155–6, 158–65, 169; *The Nurture Assumption* 160–1
Harris, Sam 147, 148, 149–50, 151, 152
Harvard Business Review 224
Harvard University 9, 28, 57, 155, 159, 300
Hayek, Friedrich 35, 102, 128, 133, 230, 232, 243; *The Constitution of Liberty* 300; *The Road to Serfdom* 253
Haynes, John Dylan 146–7
Hazlett, Tom 223
Heidegger, Martin 201
Helsinki 211, 212
Henrich, Joe 89
Henry II 34
Henry VII 240
Henry the Navigator, Prince 134
Heraclius 262
Heritage Foundation 241
Higgs, Robert 240
Hill, P.J. 235–6
Hines, Melissa 169
Hitler, Adolf 198, 201, 217, 251, 252, 253; *Mein Kampf* 252
Hobbes, Thomas 8, 12, 197–8, 243
Holdren, John 208
Holland 142
Holland, Tom, *In the Shadow of the Sword* 261–2

Holocaust 214
Hong Kong 31, 92, 97, 101, 190, 191, 233–4
Hood, Bruce 148; *The Self Illusion* 145
Horgan, John 60
Hortlund, Per 284
'How Aid Underwrites Repression in Ethiopia' (2010) 232
Howard, John 273
Hu Yaobang 212
Human Genome Project 64
Human Rights Watch 232
Hume, David 20, 21–2, 40–1, 54, 276; *Concerning Natural Religion* 39–40; *Natural History of Religion* 257
Humphrey, Nick 144, 154
Hussein, Saddam 298
Hutcheson, Francis 22, 25
Hutchinson, Allan 33
Hutton, James 17
Huxley, Aldous, *Brave New World* 167
Huxley, Julian 205, 211
Hyderabad 181

Ibsen, Henrik 249
Iceland 32
Iliad 87
Immigration Act (US, 1924) 201
Incas 86, 259
India 34, 87, 108, 125, 177–8, 181, 183, 196, 204, 206, 213, 214, 258, 259
Industrial (R)evolution 63, 104, 108, 109–10, 135, 220, 248, 254–5, 277
Infoseek (search engine) 120
Intel 223
Intergovernmental Panel on Climate Change (IPCC) 273–4
International Code of Conduct for Information Security 305

International Federation of Eugenics Organisations 202
International Monetary Fund (IMF) 286
International Telecommunications Union (ITU) 305
internet: balkanisation of the web 302–6; and bitcoin 308–12; and blockchains 306–9, 313–14; central committee of 305–6; complexity of 300–1; emergence of 299–300; individuals associated with 301–2; and politics 314–16
Internet Corporation for Assigned Names and Numbers (ICANN) 305–6
Iraq 32, 255
Ireland 213, 246
Irish Republican Army (IRA) 240
Islam 259, 260, 262–3
Islamabad 92
Islamic State 240
Israel, Paul 119
Italian city states 101
Italy 34, 247, 251
Ive, Sir Jonathan 319

Jablonka, Eva 56, 57
Jackson, Doug 309
Jacobs, Jane 92
Jagger, Bianca 211
Jainism 260
Japan, Japanese 32, 122, 125, 231, 232, 288
Jefferson, Thomas 15, 20, 114, 244
Jehovah 13, 276
Jerome, St 11
Jesus Christ 8, 9, 88, 257, 258, 263, 266
Jevons, William Stanley 63, 106
Jews 29, 142, 197, 202–3, 257
Jobs, Steve 119, 222

Johnson, Boris 166; *The Churchill Factor: How One Man Made History* 217
Johnson, Lyndon B. 206, 207, 289
Johnson, Steven Berlin 220; *Where Good Ideas Come From: The Natural History of Innovation* 127
Jones, Judge John 49, 50, 51
Jonson, Ben 15
J.P. Morgan 290
Judaism 258, 259, 260, 261, 263
Justinian, Emperor 34, 262

Kagan, Jerome 161
Kahn, Bob 301
Kalikuppam (nr Pondicherry, India) 185–6
Kammerer, Paul 56
Kant, Immanuel 8
Kauffman, Stuart 125
Kay, John 92
Kealey, Terence 134, 137, 138
Kedzie, Christopher 300
Kelly, Kevin 122, 125, 129, 131; *What Technology Wants* 120, 126
Kennedy, Gavin 25
Kennedy, John F. 206
Kenya 170, 181, 296
Keynes, John Maynard 105
Kim Il-sung 252
Kirwan, Richard 17
Kitzmiller, Tammy 49; Kitzmiller vs Dover Area School District (2005) 49–50
Klein, Richard 83
Knight, Thomas 121
Koestler, Arthur 56
Koran 8, 260–1, 261, 262
Kosslyn, Stephen 185
Kroeber, Alfred 120
Krugman, Paul 292, 293
Kryder's Law 124

Kublai Khan 87
Kurzweil, Ray 124

Lagos 182–3
Lamarck, Jean-Baptiste de 55–7
Lamb, Marion 56, 57
Lamont, Norman 295
Lane, Nick 61, 62
Lao Tzu 31, 241
Laplace, Pierre-Simon 17–18, 41
Latin America 229, 233
Laughlin, Harry 200, 202–3
Law, John 285–6
Lawson, Nigel 273, 275
leadership: China's reform 217–19;
    and economic development
    228–33; evolution of
    management 2258; giving credit
    to 215–16; Great Man theory
    216–17, 218, 222–5, 228;
    Hong Kong example 233–4;
    mosquitoes win wars 219–22
Lee, Sir Tim Berners 301
Leeds 91
Leibniz, Gottfried 12, 14, 15, 120,
    276
Lenin, V.I. 217, 250
Lessing, Doris 188
Levellers 242–3
Libet, Benjamin 146
Library of Mendel 48
life: critics of Darwin 49–52;
    culture-driven genetic evolution
    57–8; designed 39–42;
    development of the eye 44–6;
    Lamarckian view 557; natural
    selection 38–9, 42–8; organised
    complexity 44–5
Lilburne, John 242
The Limits to Growth (Club of
    Rome) 211, 212
Lincoln, Abraham 4
Lindsey, Brink 248, 318
Lisbon earthquake (1755) 14

Little Ice Age 276
Live Well Collaborative 130
Lloyd George, David 116
Locke, John 12, 20, 39, 41, 53, 67,
    143, 243, 247
Lockheed 130
Lodygin, Alexander 119
London 91, 92, 94, 121
Looksmart (search engine) 120
Lorentz, Hendrik 121
Lorenz, Edward 18
Lost City Hydrothermal Field 61
Louis XIV 101, 142
Lovelock, James 20
Lucretia, rape of 87
Lucretian heresy 10–12
Lucretius (Titus Lucretius Carus)
    7, 8–10, 12, 14, 16, 21, 52, 244,
    268; De Rerum Natura (Of the
    Nature of Things) 8–12, 13,
    15–16, 21, 37, 59, 76, 96, 118,
    140, 155, 174, 193, 215, 235,
    256, 277
Luther, Martin 8, 216
Lycos 120
Lyft 109
Lysenko, Trofim 157

M-Pesa 296
Macbook Air laptop 319
McCloskey, Deirdre 96–7, 104,
    108, 217n, 229, 248; The
    Bourgeois Virtues 32
Mackintosh, James 38
McNamara, Robert 206, 208
McNeill, J.R. 220–2
Mackey, John 227
Maccoby, Eleanor 161
Machiavelli, Niccolò 15
Madras 186
Mafia 238, 239, 240
Malawi 232–3
Malthus, Robert 38, 104, 193,
    194–7, 203, 204–5, 208,

213–14, 246; *Essay on Population* 120, 194
Manchester 91
Mandela, Nelson 217
Manhattan 91
'Mankind at the Turning Point' (Club of Rome) 211
Mann, Charles, *1493* 220
Mann, Horace 176, 189
Mansfield, Edwin 133
Mao Zedong 210, 217, 219, 252
Marconi, Guglielmo 124
Marcus Aurelius 9
Margarot, Maurice 244
Marinetti, Filippo 198
Marshall, Alfred 106
Martin, William 61
Martineau, Harriet 38, 244–5; *Illustrations of Political Economy* 244
Marx, Karl 8, 106, 165, 216n, 248, 252, 269–70
Marxism 104, 267, 302
Maude, Francis 255
Maupertuis, Pierre-Louis 14–15
Maurice, Prince of Saxe 88
Mauritius 125
Max Planck Institute, Leipzig 146–7
May, Tim 306
Mayans 259
Mead, Carver 123
Mecca 260, 261, 261402
Medawar, Sir Peter 211
Medicaid 114
Medicare 114
Men in Black (film, 1997) 141
Mencken, H.L. 189
Mendel, Gregor 121–2, 199
Menger, Carl 106
Mexico 87, 170, 238
Micklethwait, John 247
Middle Ages 88
Mill, John Stuart 104, 105, 187, 246, 247, 249

Miller, George A. 159
Miller, Kenneth 51
Milton, John 15
mind: background 140–2; and the brain 143–7; and free will 142–3, 147–54; responsibility in a world of determinism 150–4; and self 140–1
mind–body dualism 141
Minerva Academy, San Francisco 184–5
Ming Chinese 130
Mises, Ludwig von 112
Mississippi Company 286
MIT (Massachusetts Institute of Technology) 184, 301
Mitchell Energy 136
Mitchell, George 136
Mitra, Sugata 176–7, 185–7
Moglen, Eben 303
Mohamed 8, 216, 257, 260–3, 266
Molière (Jean-Baptiste Poquelin) 15
money: crypto-currencies 296, 308–9, 310–12; emergence of 277–80; fiat money 297; financial crisis 287–94; financial stability without central banks 284–6; main functions 296; mobile money 294–8; and nationalisation of system 283–4; Scottish experiment 280–2; sub-prime market 289–94
Mongols 101
Monism 197–8
Montaigne, Michel de 15
Montana 92
Montesquieu, Charles-Louis de Secondat, Baron de La Brède et de 20, 31, 142, 216; *The Spirit of the Laws* 16
Montessori schools 188
Montford, Andrew 188

Monty Python's Life of Brian (film, 1979) 42, 265
Moore, Gordon 123
Moore's Law 123–5
Morality: effect of commerce on 30–3; emergence of 26–7, 28; evolution of 28–30; impartial spectator 24–5, 30; nature-via-nurture explanation 23–4; spontaneous phenomenon 21–2, 25
More, Thomas, *Utopia* 15
Mormonism 263
Morning Star Tomatoes 225–8
Morris, Ian, *War: What is it Good For?* 239
Morris, William 248
Moses 263, 264
Mosley, Oswald 251
Mountain Meadow massacre (1857) 89
Mozart, Wolfgang Amadeus 85
Muir, Thomas 244
Mumbai 92
Murphy, Archibald 176
Muslims 52, 89, 263
Mussolini, Benito 251, 252
Myrdal, Gunman 230

Nakamoto, Dorian Satoshi 309
Napoleon Bonaparte 101, 175, 216, 280
Napoleon III 247
Nation 250
National Health Service (NHS) 116
National Institute of Child Health and Human Development 161
National Mortgage Corporation 287
Natural Theology 25
Nazis 175, 196, 198, 201, 202–3, 253, 318
Neanderthals 82, 83

Necker cube 145
'Negro Project' (1939) 201
Nelson, Richard 137, 138
neo-Malthusians 209
Neptune 120–1
New Deal 251, 290
New Delhi 185
New Guinea 80, 81
New Jersey 121
New Statesman 315
New York 92, 121, 167, 176
New York Times 291
New Zealand 32, 177
Newcastle University 181
Newcastle upon Tyne 91, 119
Newcomen, Thomas 1–2
Newton, Isaac 13, 14, 17, 20, 21, 23, 41, 51, 120, 215; *Opticks* 13
Niccolò Niccoli 12
Nietzsche, Friedrich 8
Nigeria 181
Nobel Prize 122, 230, 273
Nock, Albert Jay 240
Norberg, Johan 284
North Korea 32, 101, 102, 114
North of Scotland Bank 281
Norway 32, 247
Not in Our Genes (Lewontin, Rose & Kamin) 157
The Not So Wild, Wild West (Anderson & Hill) 236
NotHaus, Bernard von 309
Noyce, Robert 223
Nuremberg laws 198

Obama, Barack 219–20, 300
Odyssey 87
OECD (Organization for Economic Cooperation and Development) 139
Of Pandas and People (Kenyon et al.) 50
Office of Population 206
Ogburn, William 127–8

O'Grady, Selina, *Man Created God* 256, 257
Ohno, Susumu 69–70, 71
Oktar, Adnan (Harun Yahya) 52
Opium Wars 233, 245
Oppenheimer, Robert 119
Orgel, Leslie 67
Orszag, Jonathan 292
Orszag, Peter 292
Orwell, George 300
Osborn, Frederick 204
Osborn, Henry Fairfield 200, 204, 205; *Our Plundered Planet* 203–4
Otteson, James 23, 24, 26, 27
Overton, Richard 242

Paddock, William, *Famine 1975!* (with Paul Paddock) 207
Page, Larry 188
Pagel, Mark 80, 81–2
Pakistan 32, 206
Paley, William 38–9, 41–2, 51
Panama 286
Paris 102, 121, 254
Park, Walter 139
Parris, Matthew 303
Parys Mine Company, Anglesey 278
Pascal, Blaise 273
Paul, Senator Rand 241
Paul, Ron 114, 285, 292, 295
Paul, St (Saul of Tarsus) 8, 258, 264
Pauling, Linus 121
Pax Romana 239
Peace High School, Hyderabad (India) 181
Peel, Robert 246, 283–4
Peer-to-Peer Foundation 308
Peninsular War 280
People's Printing Press 288
personality: and the blank slate 156–7, 158–9; and genes 159, 160–2; and homicide 169–71; innateness of behaviour 157–8; intelligence from within 165–7; non-genetic differences 162–5; and parenting 159–60, 161–2; and sexual attraction 172–3; and sexuality 167–9
Peterloo massacre (1819) 245
Pfister, Christian 276
Philippe, duc d'Orléans 286
Philippines 190
Philips, Emo 140
Philostratus 258
Phoenicia 101
Pinker, Steven 28, 30, 31–3, 172–3; *The Better Angels of Our Nature* 28–9
Pinnacle Technologies 136
Pitt-Rivers, Augustus 127
Pixar 124
Planned Parenthood Foundation 204
Plath, Robert 126
Plato 7, 11
Plomin, Robert 165, 167
Poincaré, Henri 18, 121
Polanyi, Michael 133, 253
politics 314–16
Poor Law (1834) 195
Pope, Alexander 15
Popper, Karl 253; 'Conjectures and Refutations' 269
Population: American eugenics 200–3; control and sterilisation 205–8; and eugenics 197–9; impact of Green Revolution 208–10; Irish application of Malthusian doctrines 195–7; Malthusian theory 193, 194–5; and one-child policy 210–14; post-war eugenics 203–5
Population Crisis Committee 206
Portugal, Portuguese 134
Pottinger, Sir Harry 233

'Primer for Development' (UN, 1951) 232
Prince, Thomas 242
Pritchett, Lant 179–80; *The Rebirth of Education* 176
Procter & Gamble 130
Proudhon, Pierre-Joseph 194–5
Prussia 176
Psychological Review 159
Putin, Vladimir 305
'The Puzzle of Monogamous Marriage' (Henrich, Boyd & Richerson) 89
Pythagoras 85
Pythagorism 259

Qian XingZhong 213
Quesnay, François 98

Raines, Franklin 292
Ramsay, John 25
RAND Corporation 206, 300
Ravenholt, Reimert 206
Ray Smith, Alvy 124
Reagan, Ronald 254, 290
Red Sea 82
Reed, Leonard 43
Reformation 216, 220
religion: and climate change/global warming 271–6; and cult of cereology (crop circles) 264–6; existence of God 14–15; heretics and heresies 141–2; as human impulse 256–8; predictability of gods 259–60; and the prophet 260–3; temptations of superstition 266–8; variety of beliefs 257–8; vital delusions 268–71
Renaissance 220
Ricardo, David 104–5, 106, 246
Richardson, Samuel 88
Richerson, Pete 78, 89
Ridley, Matt, *The Rational Optimist: How Prosperity Evolves* 110–11, 126–7
Rio de Janeiro 92
Roberts, Russ 4
Robinson, James 97–8
Rockefeller Foundation 229, 230–1
Rodriguez, Joã 47–8
Rodrik, Dani 228
Rome 257, 259, 260
Romer, Paul 109
Roosevelt, Franklin Delano 251
Roosevelt, Theodore 197
Rothbard, Murray 243
Rousseau, Jean-Jacques 165, 216
Rowling, J.K. 122
Royal Bank 281
Royal Mint 278, 279
Royal Navy 297
Royal United Services Institution 198
Rudin, Ernst 202
Rufer, Chris 226
Runciman, Garry, *Very Different, But Much the Same* 94
Rusk, Dean 206–7
Russell, Lord John 195
Russia 119, 204, 227–8, 250, 303
Russian Revolution 318

Sadow, Bernard 126
Safaricom 296
St Louis (ship) 202–3
St Maaz School, Hyderabad (India) 181
Salk Institute, California 67
San Marco, Venice 53
Sandia National Laboratory 136
Sanger, Margaret 201, 204
Santa Fe Institute 93, 126
Santayana, George 10
Sapienza, Carmen 67
Satoshi Nakamoto 307–8, 309–10, 312
Schiller, Friedrich 248

Schmidt, Albrecht 222
Schumpeter, Joseph 106, 128, 251; *Capitalism, Socialism and Democracy* 106–7; *Theory of Economic Development* 106
science: as driver of innovation 133–7; as private good 137–9; pseudo-science 269
Science (journal) 70
Scientology 263
Scopes, John 49
Scotland 17, 280–2, 286
Scott, Sir Peter 211
Scott, Sir Walter ('Malachi Malagrowther') 283
Second International Congress of Eugenics 200
Second World War 105, 138, 203, 231, 252, 254, 318
Self-Management Institute 226
Selgin, George 297; *Good Money* 279, 280
Shade, John 188
Shakespeare, William 15, 131, 216, 224
Shanker, Albert 180
Shaw, George Bernard 197
Shaw, Marilyn 155–6
Shelley, Mary, *Frankenstein* 16
Shelley, Percy 16
Shockley, William 119
Shogun Japanese 130
Sierra Club 204
Silk Road 311–12
Silvester, David 274
Simon, Julian 209
Singapore 190
Sistine Chapel, Rome 256
Skarbek, David, *The Social Order of the Underworld* 237–8
Skinner, B.F. 156, 267–8
Skirving, William 244
skyhooks 7, 13, 14, 18, 65, 67, 71, 150, 267

Slumdog Millionaire (film, 2008) 185
Smith, Adam 3, 20, 21, 22–7, 28, 33, 110, 112, 117, 234, 243, 244, 246, 249; *The Theory of Moral Sentiments* 23–4, 27, 28, 37–8, 98; *The Wealth of Nations* 24, 38, 98–100, 103–4, 137
Smith, John Maynard 53
Smith, Joseph 263, 264, 266
Smithism 110
Snowden, Edward 303
SOLE (self-organised learning environment) 186
Solow, Robert 108, 137
Somalia 32
Song, Chinese dynasty 101
Song Jian 210–11, 212–13
South America 247
South Korea 125, 190, 229
South Sea Bubble (1720) 285, 294
South Sudan 32
Soviet-Harvard illusion 3
Soviet Union 114, 122
Spain 101, 247
Sparkes, Matthew 313
Sparta 101
Spencer, Herbert 216–17, 249, 253
Spenser, Edmund 15
Spinoza, Baruch 20, 141–2, 148, 268; *Ethics* 142; *l'Esprit des lois* 142–3
Sputnik 138
Stalin, Joseph 250, 252, 253
Stalling, A.E. 10
Stanford University 184, 185
Stealth bomber 130
Steiner, George, *Nostalgia for the Absolute* 266
Steiner, Rudolf 271
Steinsberger, Nick 136
Stephenson, George 119
Stewart, Dugald 38, 244

Stiglitz, Joseph 292
Stockman, David 288, 289–90;
    The Great Deformation 294
stoicism 259
Stop Online Piracy Act (US, 2011)
    304
Strawson, Galen 140
Stuart, Charles Edward 'The Young
    Pretender' 282
Stuart, James Edward 'The Old
    Pretender' 281
Sudan 32
Summers, Larry 110
Sunnis 262
Suomi, Stephen 161
Sveikauskas, Leo 139
Swan, Joseph 119
Sweden 101, 284
Switzerland 32, 190, 247, 254
Sybaris 93
Syria 32
Szabo, Nick 307, 310; 'Shelling
    Out: The Origins of Money'
    307

Tabarrok, Alex 132; *Launching the
    Innovation Renaissance* 132
Taiwan 190
Tajikistan 305
Taleb, Nassim 3, 92, 107, 135, 285,
    312
Tamerlane the Great 87
Taoism 259, 260
Taylor, Winslow 250
Taylorism 250, 251
Tea Act (UK, 1773) 282n
Tea Party 246
technology: biological similarities
    126–31; boat analogy 128;
    computers 123–5, 126;
    copying 132–3; electric
    light 1–2; and fracking 136;
    inexorable progress 122–6,
    130–1; innovation as emergent

phenomenon 139; and the
    internet 299–316; light bulbs
    118–19, 120; many-to-many
    300; mass-communication 200;
    open innovation 130; patents/
    copyright laws 131–2; and
    printing 220; and science 133–9;
    simultaneous discovery 120–2;
    skunk works 130; software 131
TED (Technology, Entertainment,
    Design) lecture 177
Thatcher, Margaret 217
Third International Congress of
    Eugenics 201–2, 204
Third World 231–2
Thrun, Sebastian 185
Time (magazine) 241
The Times 308
Togo 94
Tokyo 92
Tolstoy, Leo 217
Tooby, John 43
Tooley, James 181–4
Toy Story (film, 1995) 124
Trevelyan, Charles 195
Tuchman, Barbara, *A Distant
    Mirror* 29
Tucker, William 90; *Marriage and
    Civilization* 89
Tullock, Gordon 35
Turner, Ted 213
Twister (messaging system) 313
Twitter 310, 313

U-2 reconnaissance plane 130
Uber 109
UK Meteorological Office 275
UN Codex Alimentarius 254
UN Family Planning Agency 213
UN Framework Convention on
    Climate Change 254–5
UN General Assembly 305
UNESCO 205
Union Bank of Scotland 281

United Nations 131, 213, 232, 305
United States 34, 122, 125, 138, 139, 176, 200–2, 232, 235–8, 245, 247, 250, 254, 284–5, 286, 302
United States Supreme Court 50
universe: anthropic principle 18–20; designed and planned 7–10; deterministic view 17–18; Lucretian heresy 10–12; Newton's nudge 12–13; swerve 14–15
University of Czernowitz 106
University of Houston 71
University of Pennsylvania 133
UNIX 302
Urbain Le Verrier 120–1
US Bureau of Land Management 240
US Department of Education 240
US Department of Homeland Security 240, 241
US Federal Reserve 285, 286, 288, 293, 297, 309
US Financial Crisis Inquiry Commission 294
US Internal Revenue Service (IRS) 240
US National Oceanographic and Atmospheric Administration 240
US Office of Management and Budget 290
Utah 89
Uzbekistan 305

Vancouver 92
Vanuatu 81
Vardanes, King 258
Veblen, Thorstein 249
Verdi, Giuseppe: *Aida* 248; *Rigoletto* 248
Veronica (search engine) 120
Versailles Treaty (1919) 318

Victoria, Queen 89
Virgil (Publius Vergilius Maro) 10, 23
vitalism 270–1
Vodafone 296
Vogt, William 205, 209; *Road to Survival* 204
Voltaire, François-Marie Arouet 14, 15, 20, 22, 25, 41, 143, 243, 268; *Candide* 15
Volvo 101

Wagner, Andreas 47
Wall Street Journal 125, 132
Wallace, Alfred Russell 20, 54–5, 196
Wallison, Peter 294
Walras, Léon 106
Waltham, David, *Lucky Planet* 19
Walwyn, Thomas 242
Wang Mang, Emperor 267
Wang Zhen 212
Wannsee conference 198
Wapinski, Norm 136
Washington, George 220, 222, 240
Washington Post 241
Watson, James 121, 145
Webb, Beatrice 197
Webb, Richard 5, 319
Webb, Sidney 197
Webcrawler 120
Wedgwood family 38
Wedgwood, Josiah 199
Weismann, August 55
Wells, H.G. 197, 251
West, Edwin 178; *Education and the State* 177
West, Geoffrey 93
West Indies 134, 286
Whitney, Eli 128
Whittle, Frank 119
Whole Foods 227
Wikipedia 188, 304–5
Wilby, Peter 315

Wilhelm II, Kaiser 198, 247
Wilkins, Maurice 121
Wilkinson, John 278–9
Willeys 278–9, 280
Williams, Thomas 278
Williamson, Kevin 33; *The End is Near and it's Going to be Awesome* 238–9
Wilson, Catherine 12
Wilson, Margo 171
Wolf, Alison, *Does Education Matter?* 189–91
Wolfe, Tom 223
Wooldridge, Adrian 247
World Bank 181, 183, 189–90, 206, 208, 232
World Economic Forum 224
World Forum for the Harmonisation of Vehicle Regulations 254

World Wildlife Fund 204
Wright, Chris 136

Xiaogang 218

Yahoo 120
Yale University 236
Yen Jingchang 218
YouTube 315

Zappos 227
Zenawi, Meles 232
Zeus 257, 276
Zola, Emile 249
Zoroaster 257
Zoroastrians 258, 261
Zubrin, Robert 195–6; *Merchants of Despair* 196
Zuckerberg, Mark 222, 223